Ecology of Weeds and Invasive Plants: Relationship to Agriculture and Natural Resource Management

Ecology of Weeds and Invasive Plants: Relationship to Agriculture and Natural Resource Management

Editor

Arun Kumar Das

scitus academics

Ecology of Weeds and Invasive Plants: Relationship to Agriculture and Natural Resource Management

Edited by **Arun Kumar Das**

Printed in 2017

ISBN: 978-1-68117-058-9

Library of Congress Control Number: 2015935456

SCITUS Academics LLC,
616, Corporate Way, Suite 2, 4766,
Valley Cottage, NY 10989

www.scitusacademics.com

Notice

Contents

Preface

This book provides in-depth understanding of how weeds and invasive plants develop and interact in the environment so one can manage and control them more effectively. The guide includes an introduction to weeds and invasive plants in various environments and an overview of their ecology and evolution. It focuses on the biological features of weeds and invasive plants, especially as they exist in agriculture, forests, rangelands, and natural ecosystems also includes coverage of exotic invasive plants. It discusses a variety of methods and tools for managing weeds and invasive plants, including physical, cultural, biological, and chemical approaches. And examines systems approaches for management, including modern Integrated Pest Management. Addresses future challenges for scientists, farmers, and land managers. This is the definitive, hands-on reference for a land manager or professional in plant sciences, agronomy, weed science, and horticulture. The book is also an excellent textbook for senior undergraduate or graduate students studying agriculture, ecology, natural resources management, environmental management, or related fields.

Editor

were placed by species into paper bags and dried at 72°C for 3 days to obtain shoot biomass. Cover crop and weed species shoot biomass data were converted to relative yield (RY) according to the following equations:

$$RY_{\text{crop species}} = \frac{\text{yield of weed species in mixture}}{\text{yield of weed species in monoculture}},$$
$$RY_{\text{weed species}} = \frac{\text{yield of weed species in mixture}}{\text{yield of weed species in monoculture}}, \quad (1)$$

Regression of RYs on weed proportions 0, 0.25, 0.5, 0.75, and 1.0 were used to produce the replacement series diagrams to determine the competitiveness in the mixture as compared with the monoculture [18, 22, 23]. The shape of the replacement curve of the RY for the shoot dry weight relative to expected yields was used as the indicator of the extent of interference between the two competing species [24].

Relative yield totals (RYTs), which predict the competition between the two species for the same resources, were calculated as described by Santos et al. [25] by using the following equation:

$$RYT = RY\text{crop species} + RY\text{weed species}. \quad (2)$$

The relative crowding coefficient (RCC), which serves as an index of competition when two species are mixed in equal proportions, was determined using the following equation [20]:

$$RCC = \frac{W1m / W2m}{W1p / W2p} \text{ or } \frac{W2m / W1m}{W2p / W1p}, \quad (3)$$

Where $W1m$ and $W2m$ are shoot dry weight per pot of crop species and weed species at C_{50}: W_{50} mixture and $W1p$, and $W2p$ are shoot dry weight per pot of crop species and weed species in pure culture (monoculture). Equivalent yield ratios (EYR) or the proportion at which both species growing in the mixture produce the same yield was calculated for each mixture [25].

Statistical Analysis

Plant height and leaf area data were analysed by the analysis variance using SAS statistical software package version 9.2 [26], and values were further differentiated by Tukey's test at $P \leq 0.05$. All regressions were conducted using Sigma Plot version 11.

RESULTS

Competitive Ability of A. compressus against A. gangetica in Full Sunlight

Since density by proportion interaction was significant for shoot dry weight of A. compressus, every combination was analyzed separately (data not given). A. compressus in monoculture (C_{100}), C_{75}:W_{25} and C_{50}:W_{50} produced less shoot biomass per plant than in C_{25}:W_{75} at two different densities (Table 1). Moreover, the highest shoot dry weights plant^{-1} across the different proportions was found at 288 plants m^{-2}. The mean shoot dry weight per plant of A. gangetica decreased as the proportions of A. compressus increased. The highest A. gangetica shoot dry weight was obtained in the pure stand of A. gangetica (W_{100}) and the lowest in $C_{75}W_{25}$ and $C_{50}W_{50}$ at 288 plants m^{-2}. At 72 plants m^{-2} density, the lowest A. gangetica shoot dry weight was in $C_{75}W_{25}$ (Table 1), and other proportions did not show any significant difference from the pure stand.

Table 1: Shoot dry weight, plant height, and leaf area of A. compressus and A. gangetica in monoculture and mixtures at 10 wk after planting in full sunlight

Treatments	Shoot dry weight plant–1 (g)		Plant height (cm)		Leaf area plant–1 (cm2)	
	A. compressus	A. gangetica	A. compressus	A. gangetica	A. compressus	A. gangetica
72 plants m–2						
C100	3.1b	—	4.5a	—	244.6a	—
C75W25	2.8b	0.8b	5.5a	17.0b	238.7a	32.0b
C50W50	4.2b	1.9a	4.8a	20.0b	355.3a	123.3a
C25W75	6.6a	1.0ab	3.9a	15.0b	479.9a	243.3a
W100	—	2.1a	—	27.7a	—	142.8a

288 plants m−2						
C100	10.4b	—	5.0a	—	613.3b	—
C75W25	9.7b	0.02c	4.5a	7.0b	601.6b	2.3c
C50W50	10.4b	0.12c	4.5a	6.8b	636.0b	13.7c
C25W75	16.4a	0.48b	5.0a	6.2b	909.3a	40.5b
W100	—	1.99a	—	20.9a	—	134.3a

Means within column for each density followed by the same letter are not significantly different at $P \leq 0.05$. %.C: A. compressus, W: A. gangetica.

Plant height of A. compressus was unaffected in different proportions. A. gangetica had the highest plant height in monoculture and a reduced height in mixtures. In General, plant heights of A. compressus were lower thanA. gangetica in monoculture (Table 1). Leaf areas of A. compressus in association with A. gangetica showed a similar trend to the shoot dry weight at 288 plants m^{-2} density as A. compressus had the highest leaf area in $C_{25}W_{75}$. Although A. compressus showed this response at 72 plants m^{-2} density, it was not significant (Table1). A. gangetica leaf area decreased with increasing A. compressus proportions in the mixture as A. gangeticahad the lowest leaf area in $C_{75}W_{25}$ at 72 plants m^{-2} and in $C_{75}W_{25}$ and $C_{50}W_{50}$ at 288 plants m^{-2} (Table 1).

The De Wit competitiveness diagrams of the relative shoot dry weight (RY) of the A. compressus and A. gangetica is shown in Figures 2(a) and 2(b). The RY of A. compressus (Figure 2(a)) increased in a quadratic manner as the proportion of it in mixtures with A. gangetica increased, resulting in a convex curve at 72 plants m^{-2}. As the proportion of A. gangetica in the mixtures increased, the RY of A. compressus decreased in a linear manner and near to the expected curve at 288 plants m^{-2} (Figure 2(b)). A. gangetica responded to A. compressus to form a concave curve at both densities (Figures 2(a) and 2(b)) that resulted in an equivalent yield ratio (EYR) of more than 0.50. The EYRs were 0.68 and 0.75 for A. compressus at 72 and 288 plants m^{-2}, respectively (Figures 2(a) and 2(b)). The RYT of mixture was equal to monoculture at 72 plants m^{-2} (Figure2(a)). The RYT value was less than 1 at 280 plants m^{-2} (Figure 2(b)). The Relative crowding coefficient (RCC) values of A. compressus at both densities, when grown in equal proportions, were more than the RCCs of A. gangetica (Table 3).

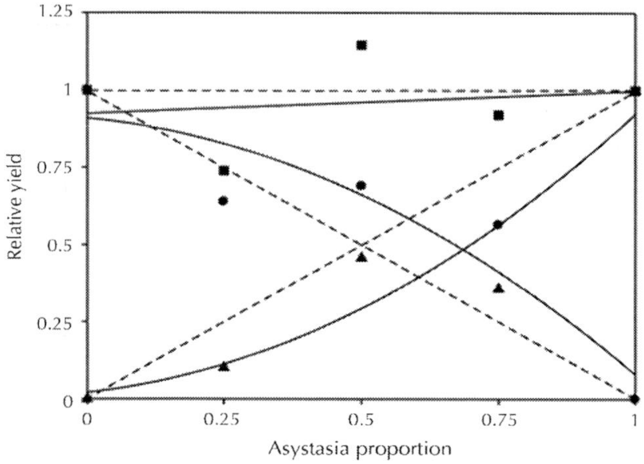

● Axo
▲ Asy
■ T

Axo: $Y = -0.66X^2 - 0.66X + 0.91$, $R^2 = 0.86$

Asy: $Y = 0.71X^2 + 0.18X + 0.02$, $R^2 = 0.88$

(a)

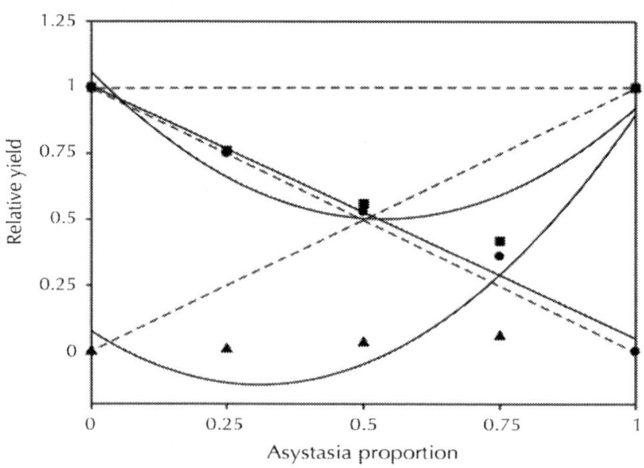

● Axo
▲ Asy
■ T

Axo: $Y = -0.95X + 1.006$, R2 = 0.98

Asy: $Y = 2.13X^2 - 1.31X + 0.07$, $R^2 = 0.87$

(b)

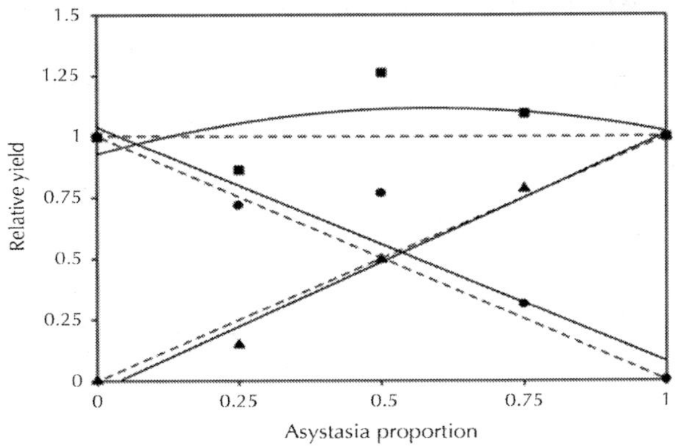

● Axo
▲ Asy
■ T

Axo: Y = -0.96X + 1.04, R^2 = 0.91
Asy: Y = 1.05X - 0.04, R^2 = 0.98

(c)

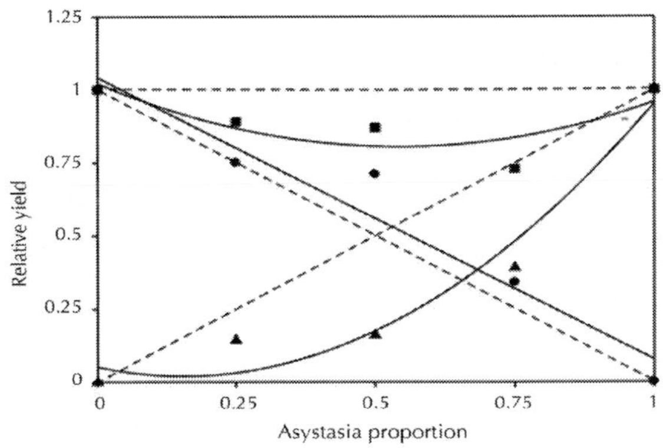

● Axo
▲ Asy
■ T

Axo: Y = -0.96X + 1.04, R^2 = 0.94
Asy: $1.3X^2$ - 0.4X + 0.05, R^2 = 0.95

(d)

Figure 2: Interaction between Asystasi gangetica (Asy) and Axonopus compressus (Axo) in full sunlight with low density (a), A. compressus in full sunlight with high density (b), A. compressus in shade with low density (c), and A. compressus in shade with high density (d) in replacement series experiments at 10 wk after planting. T = total.

Competitive Ability of A. compressus against A. gangetica in Shade (Glasshouse Condition)

Shoot biomass of A. compressus at 72 plants m^{-2} did not show a significant difference amongst the different proportions, but at 288 plants m^{-2} density the A. compressus in $C_{25}W_{75}$ and $C_{50}W_{50}$ produced more shoot biomass per plant than in monoculture (C_{100}) and in $C_{75}W_{25}$ (Table 2). A. gangetica produced the highest shoot biomass per plant in $C_{25}W_{75}$ proportion in both densities. However, A. gangetica produced the lowest shoot biomass in $C_{75}W_{25}$ at 72 plants m^{-2} and in $C_{75}W_{25}$ and $C_{50}W_{50}$ at 288 plants m^{-2} (Table 2).

Table 2: Shoot dry weight, plant height, and leaf area of A. compressus and A. gangetica in monoculture and mixtures at 10 wk after planting in shade

Treatments	Shoot dry weight plant−1 (g)		Plant height (cm)		Leaf area plant−1 (cm2)	
	A. compr-essus	A. gange-tica	A. comp-ressus	A. gang-etica	A. compr-essus	A. ange-tica
72 plants m−2						
C100	3.3a	—	15.0a	—	201.9a	—
C75W25	3.4a	0.21c	15.0a	19.3a	204.6a	29.2c
C50W50	5.1a	1.07b	13.0c	21.8a	307.7a	221.4b
C25W75	4.1a	2.40a	14.0b	18.1a	254.0a	436.3a
W100	—	1.08b	—	23.9a	—	144.5b
288 plants m−2						
C100	7.4b	—	19.0a	—	477.3b	—
C75W25	6.8b	0.20c	19.0a	21.8a	499.2b	20.9d
C50W50	9.6a	0.33c	14.6b	19.0b	703.7a	47.0c
C25W75	9.3a	1.66a	14.6b	20.8ab	660.8a	163.5a
W100	—	1.03b	—	21.3ab	—	133.1b

Means within column for each density followed by the same letter are not significantly different at %.

C: A. compressus, W: A. gangetica.

Table 3: Relative crowding coefficient in interaction between A. compressus and A. gangetica

Treatments	Relative crowding coefficient	Relative crowding coefficient
	A. compressus/A. gangetica	A. gangetica/A. compressus
In full sunlight		
72 plants m–2	1.51	0.66
288 plants m–2	4.04	0.24
In shade		
72 plants m–2	1.01	0.90
288 plants m–2	1.54	0.64

With the increasing proportion of A. gangetica in mixtures, plant height of A. compressus decreased at both densities (Table 2). Plant height of A. gangetica was unaffected at 72 plants m^{-2} but decreased with increasing proportion of A. compressus in mixtures at 288 plants m^{-2}. A. compressus leaf area did not show any difference between the different proportions at 72 plants m^{-2} and had the biggest leaf area in $C_{25}W_{75}$ and $C_{50}W_{50}$, at 288 plants m^{-2} (Table 2). Leaf area of A. gangetica decreased with increasing A. compressus proportions in the mixture as A. gangetica had the lowest leaf area in $C_{75}W_{25}$ at both densities. A. gangetica had the highest leaf area in $C_{25}W_{75}$ at both densities compared with monoculture at both densities (Table 2).

Replacement series curves of interaction of A. gangetica with A. compressus in the shade have been illustrated in Figures 2(c) and 2(d). The RY of A. compressus increased in a linear manner and more than expected as the proportion of it in mixtures with A. gangetica increased at both densities. The response of A. gangetica to A. compressus with normal density under the shade was linear and resulted in a 0.56 EYR (Figure 2(c)). A. gangetica responded to A. compressus to form a concave curve at 288 plants m^{-2} density (Figure 2(d)) that resulted in an EYR of about 0.68. The RYT value at 72 plants m^{-2} was greater than 1.0 (Figure 2(c)). The RYT value was less than 1.0 at 280 plants m^{-2} (Figure 2(d)). The RCC values of A. compressus in both densities, when grown in equal proportions, were more than RCCs of A. gangetica on them (Table 3).

DISCUSSION

A. compressus shoot dry weight increased despite a decreasing number of A. compressus when grown with A. gangetica compared to monoculture at 72 and 288 plants m^{-2} in the open and at 288 plants m^{-2} in the shad. It appears that A. compressus produces more dry weight per plant when competing against A. gangetica than in monoculture. A. compressus shoot dry weight remained unchanged from $C_{100}W_0$ to $C_{50}W_{50}$; hence, as proportions changed in this range both intra- and interspecific competition were counteracting each other. However, as fewer A. compressus plants were in the mixture ($C_{25}W_{75}$), neighbouring plants apparently did not compete with each other, resulting in no intraspecific competition. Under these conditions, A. compressus was more efficient in competing against A. gangetica than against other A. compressus plants. By contrast, shoot dry weight of A. gangetica decreased with increasing A. compressus proportions in the mixture as A. gangetica had the lowest shoot dry weight in $C_{75}W_{25}$ at both densities in the open and shade. These findings suggest that A. compressus responded plastically to competition, whereas A. gangetica did not. The greater biomass of A. compressus compared with A. gangetica in a mixture would result in a greater demand for resources. Other studies have also shown that more competitive species produce a higher relative yield when grown in mixtures, whereas the yield of weak competitors is lower in mixtures than in monoculture [27, 28]. When one plant of basil (Ocimum sanctum) was competing with three of weed species, plant height and fresh weight plant^{-1} of basil increased [29]. Overyielding has been associated with higher biomass density and light interception or greater demand for resources [28, 30, 31].

A. gangetica at both densities in the open grew better in monoculture compared to C:W mixtures, indicatingAsystasia gangetica was more affected by interspecific interactions. By contrast, A. gangetica at both densities in the shade in $C_{25}W_{75}$ proportion had more shoot dry weight relative to monoculture. It seems that A. gangetica plants in the shade in $C_{25}W_{75}$ did not compete with each other in this proportion and were more competitive against A. compressus which helped to neutralize additional interspecific completion by A. compressus in this proportion. Therefore, A. gangetica grew better in the mixture than in monoculture and was a better competitor than the A. compressus in the shade compared to the open. Moreover, A. compressus did not over yield at 72 plants m^{-2} in the shade at $C_{25}W_7$. It appears that A. compressus was less efficient in competing against the A. gangetica in low density.

Plant height of A. gangetica was reduced in different proportions in the open. Generally, plant heights of A. compressus were lower than A. gangetica in monoculture in the open, but unaffected by competition. Plant height of A. compressus decreased at both densities in the shade. Plant height of A. gangetica decreased with increasing proportion of A. compressus in mixtures at 288 plants m^{-2} but was unaffected at 72 plants m^{-2} in the shade. Variation between plant heights of A. compressus and A. gangetica in the shade was lower than in the full sunlight, because in the shade they were trying to achieve more light for growth. A. compressus and A. gangetica are shade tolerant [9, 14]. Shading has been shown to drastically reduce plant growth [32]. Despite this fact, A. gangetica canopy height in the open, which was taller than A. compressus, did not influence A. compressus height, but A. gangetica canopy shade could effect on height of A. compressus and reduce that in both densities in the shade.

Leaf areas of A. compressus in association with A. gangetica also responded in a similar way to shoot dry weight at 72 and 288 plants m^{-2} in the open and at 288 plants m^{-2} density in the shade. Thus, A. compressus had the biggest leaf area in $C_{25}W_{75}$. Greater leaf area expansion rates would favour A. compressus plants in competition for light. By contrast, leaf area of A. gangetica decreased with increasing A. compressusproportions in the mixture as A. gangetica had the lowest leaf area in $C_{75}W_{25}$ at both densities in the open and shade. Plant size suggests a potential advantage for light capture and greater penetration of PAR to the soil surface [33].

As the proportion of A. gangetica in the mixtures increased in open, A. compressus yielded over the expected rate and produced a convex curve at 72 plants m^{-2} and a linear response with more than expected at 288 plants m^{-2}, whereas A. gangetica yielded under the expected rate and had a concave curve at both densities in the open. Response of A. gangetica to A. compressus with normal density (72 plants m^{-2}) under the shade was linear and less than expected, and Asystasia gangetica responded at 288 plants m^{-2} density to A. compressusunder shade to form a concave curve. A convex curve for one species and a concave curve for the other species in the series indicate that the species are competing for a common resource. When both curves are convex and concave, mutually stimulatory and antagonistic relations are indicated, respectively [24]. The RY of A. compressus increased in a quadratic manner or linearly to more than expected, while RY of A. gangetica in interaction with A. compressus was concave or linear but less than expected in all conditions, indicating that A. gangetica was more affected by interspecific interactions with A. compressus and was less competitive than A. compressus.

If the RY curves intersect at 50:50 proportions, the two competing species are relatively equal in competitiveness [25]. The RY of A. gangetica increased in a linear or nonlinear manner as its proportion in the mixture with A. compressus increased, but its RY was not equivalent to that of A. compressus when each comprised half the mixture. This resulted in EYR to equal to 0.75 at 288 plants m^{-2} in the open. This meant that about one A. compressus plant equaled the shoot dry weight production of three A. gangetica plants compared to the total shoot dry weight production of each monoculture, and suggesting that a large population of A. gangetica is needed to suppress a smaller population of A. compressus. The EYR for A. compressus at 72 in the open was 0.68. Response of A. gangetica to A. compressus at 72 and 288 plants m^{-2} in the shade resulted in a 0.56 and 0.68 EYR, respectively. Axonopus compressus seemed to be more competitive than A. gangetica at 288 plants m^{-2} in the open compared to other conditions, and A. compressus was not a good competitor at 72 plants m^{-2} in the shade. For EYR, the A. compressus performed in the following order:A. compressus at 288 plants m^{-2} in the open > A. compressus at 72 plants m^{-2} in the open and A. compressus at 288 plants m^{-2} in the shade > A. compressus at 72 in the shade.

An RYT value around 1 indicates that the same resource or area is being used by the two competing species (overlap in resource utilization) [20]. An RYT > 1 indicates some niche differentiation between the species, where competition is either avoided or minimized [17]. However, other processes can also produce RYT > 1, indicating facilitation where one species benefits another [34]. An RYT < 1 suggests mutual antagonism. The RYT value was less than 1 at 280 plants m^{-2} in the open and shade, and 72 plants m^{-2} in the open means that mutual antagonism is occurring with the species producing less than expected when grown together [35]. Anyway, RYT reduced in the following order: 280 plants m^{-2} in the open > 280 plants m^{-2} in the shade > 72 plants m^{-2} in the open. It is likely allelopathy to be involved in the interaction of A. compressus and A. gangetica at 280 plants m^{-2} in the open and shade, because the RYT value of A. compressus with A. gangeticawas < 1 for all proportions. The occurrence of allelopathic interaction would have lowered the total yield in mixtures compared with monoculture [16]. The RYT value at 72 plants m^{-2} in the shade was greater than 1.0 suggesting that the two species made different demands on resources leading to better growth of A. compressusor that this crop mixture was less affected by interspecific competition than by intraspecific competition, facilitating over yielding (RYT >1).

The relative crowding coefficient (RCC) value demonstrates the aggressiveness of one species towards another. The greater (RCC) values of A. compressus than the RCCs of A. gangetica at both densities in the open and shade, when grown in equal proportions, confirms the aggressiveness of A. compressus against the A. gangeticain terms of shoot dry weight production. When competing for limited resources, the species with the greater RCC in the mixture is the strongest competitor [17].

The superior competitiveness of A. compressus relative to A. gangetica is likely resource competition. Axonopus compressus became extremely dense fast, thereby limiting the space available to the weed population and suppressing A. gangetica growth. Fast growth result in a superior plant [36]. Furthermore, plant size suggests a potential advantage for light capture and greater penetration of PAR to the soil surface, making the crop less competitive against weeds [33]. Also, rapid growth by lateral spread of A. compressus through tillering seemed to be the reason for superiority of A. compressus in competition with A. gangetica. Greater tiller production is one of the factors associated with superior suppression [37]. The aggressiveness of A. compressus can also be explained in terms of its prolific rooting system, which enabled it to capture more of the limited soil water and nutrients and resulted in rapid growth in terms of biomass accumulation and canopy development [38]. Moreover, the presence of allelopathic interaction would have lowered the total yield in mixtures compared with the monocultures [16]. Axonopus compressus is known to produce allelochemicals that affect the growth of other plants [39]. There are some reports that demonstrate A. compressus had competitive ability. Oka Nurjaya [21] reported that in mixtures, A. compressus was more competitive than grass and legume species. Rika et al. [15] observed that under coconut, a local cultivar of A. compressus produced higher yields than other grasses. Axonopus sp., Digitaria sp., and Palspalum sp. are classified as soft weeds in oil palm plantation which maintain the balance of the weed flora and prevent weed succession by noxious species simply because the base land for the noxious weeds to colonise is less available [4, 40]. On most oil palm plantations in the far East, the cover that establishes itself is a mixture of the fern Nephrolepis biserrata with varying components of grasses such Paspalum conjugatum and Axonopus compressus [41, 42].

Based on the present findings, it can be concluded that A. compressus is highly competitive against A. gangetica. However, A. compressus under shade uses most of its energy to achieve more light for growth, and hence, Axonopus compressus density should be higher under shade to increase its competitiveness with A. gangetica. Further research into the A. compressus competitive ability with weeds is planned.

ACKNOWLEDGMENTS

The authors greatly appreciate the personnel at the Department of Crop Science and the Universiti Agriculture Park (UAP), and Universiti Putra Malaysia (UPM), for their assistance.

REFERENCES

1. A. C. Soh, T. Y. Wong, C. C. Hor, P. S. Tan, and P. S. Chew, "Oil palm," in Oil Crops, J. Vollmann and I. Rajcan, Eds., pp. 333–367, Springer, New York, NY, USA, 2010.

2. D. Pride, "Environmentally friendly weed control in oil palm plantations," 2012,
3. K. H. Yeow, T. K. Tam, and M. Hashim, "Effects of interline vegetation and management on oil palm performance," in The Oil Palm Agriculture in the Eighties, E. Pushparajah and C. P. Soon, Eds., pp. 277–288, Malaysian Palm Oil Board, Kuala Lumpur, Malaysia, 1982.
4. C. H. Lam, J. K. Lim, and J. Badrulisan, "Comparative studies of paraquat mixtures and glyphosate and or its mixtures on weed succession in plantation crops," The Planter, vol. 69, pp. 525–535, 1993.
5. W. Wibawa, R. Mohamad, A. S. Juraimi, D. Omar, M. Ghazali Mohayidin, and M. Begum, "Weed control efficacy and short term weed dynamic impact of three non-selective herbicides in immature oil palm plantation," International Journal of Agriculture and Biology, vol. 11, no. 2, pp. 145–150, 2009.
6. T. Kustyanti and P. Horne, "The effect of Asystasia on the growth of young rubber in polybags," 1991,
7. C. H. Teoh, P. Y. Toh, and H. Khairuddin, "Chemical control of Asystasia intrusa (Bl), Clidemia hirta(Don.) and Elettariopsis curtisii (Bak.) in rubber and oil palm plantations," in Protection in Tropics, pp. 497–510, 1982.
8. R. Abdullah, "Biological control of Asystasia by sheep grazing," Planters' Bulletin, vol. 183, pp. 43–49, 1985.
9. K. H. Ong, M. T. Lim, P. Priscilla, and C. John Keen, "Ground vegetation response to fertilization in anAzadirachta excelsa stand in Johore, Malaysia," Journal of Agronomy, vol. 7, no. 4, pp. 327–331, 2008.
10. MADI, "Major agricultural commodities," in Malaysia Agricultural Directory and Index, Agriguuest and UMAGA, Kuala Lampur, Malaysia, 2009-2010.
11. K. J. Goh and S. B. Chiu, Mucuna Bracteata. A Cover Crop Living Green Manure, Agricultural Crop Trust (ACT), Petaling Jaya, Malaysia, 2007.
12. R. H. V. Corley and P. B. Tinker, The Oil Palm, Blackwell Science, 4th edition, 2003.

13. P. D. Turner and R. A. Gillbanks, Oil Palm Cultivation and Management, Society of Planters, 2nd edition, 2003.
14. A. S. Jurami, "Turf grass: types, uses and maintenance," Garden Asia, vol. 8, pp. 40–43, 2003.
15. I. K. Rika, et al., "New forage species for coconut plantation in Bali," in Proceedings of the Workshop on Forage for Plantation Crops, H. M. Shelto and W. W. Stur, Eds., pp. 41–43, ACIAR, Canberra, Australia, 1990.
16. H. T. Cralle, T. B. Fojtasek, K. H. Carson et al., "Wheat and Italian ryegrass (Lolium multiflorum) competition as affected by phosphorus nutrition," Weed Science, vol. 51, no. 3, pp. 425–429, 2003.
17. A. J. Fischer, C. G. Messersmith, J. D. Nalewaja, and M. E. Duysen, "Interference between spring cereals and Kochia scoparia related to environment and photosynthetic pathways," Agronomy Journal, vol. 92, no. 1, pp. 173–181, 2000.
18. C. T. de Wit, "On competition," Versal. Landbouwk. Onderz. Ned., vol. 66, pp. 1–82, 1960.
19. R. L. Zimdahl, Fundamentals of Weed Science, Elsevier Inc., 2007.
20. P. A. Jolliffe, A. N. Minjas, andV. C. Runeckles, "A reinterpretation of yield relationships in replacement series experiments," Journal of Applied Ecology, vol. 21, no. 1, pp. 227–243, 1984.
21. I. M. Oka Nurjaya, "Competition between some grass and legume species with Axonopus compressusunder reduced light intensity," Jurnal Biologi, 2006, .
22. R. Cousens, "Aspects of the design and interpretation of competition (interference) experiments," Weed Technology, vol. 5, pp. 664–673, 1991.
23. A. Weigelt and P. Jolliffe, "Indices of plant competition," Journal of Ecology, vol. 91, no. 5, pp. 707–720, 2003.
24. J. L. Harper, Population Biology of Plants, Academic Press, London, UK, 1977.
25. B. M. Santos, T. A. Bewick, W. M. Stall, and D. G. Shilling, "Competitive interactions of tomato (Lycopersicon esculentum) and nutsedges (Cyperus spp.)," Weed Science, vol. 45, no. 2, pp. 229–233, 1997.

26. SAS, The SAS System for Windows, Version 9.2, SAS Institute Inc., Cary, NC, USA, 2004.
27. J. M. Lee and M. D. K. Owen, "Dry matter yield differences of five common cocklebur (Xanthium strumarium) biotypes grown at a common site," Weed Science, vol. 51, no. 2, pp. 186–190, 2003.
28. Y. Z. Gao, S. P. Wang, X. G. Han, B. D. Patton, and P. E. Nyren, "Competition between Artemisia frigidaand Cleistogenes squarrosa under different clipping intensities in replacement series mixtures at different nitrogen levels," Grass and Forage Science, vol. 60, no. 2, pp. 119–127, 2005.
29. S. Singh and M. Singh, "Interactions of basil (Ocimum sanctum L.) with some weed species- competition or allelopathy?" Indian Journal of Weed Science, vol. 41, pp. 12–22, 2009.
30. E. Vojtech, M. Loreau, S. Yachi, E. M. Spehn, and A. Hector, "Light partitioning in experimental grass communities," Oikos, vol. 117, no. 9, pp. 1351–1361, 2008.
31. E. M. Spehn, J. Joshi, B. Schmid, M. Diemer, and C. Körner, "Above-ground resource use increases with plant species richness in experimental grassland ecosystems," Functional Ecology, vol. 14, no. 3, pp. 326–337, 2000.
32. S. Benvenuti, M. Macchia, and A. Stefani, "Effects of shade on reproduction and some morphological characteristics of Abutilon theophrasti Medicus, Datura stramonium L. and Sorghum halepense L. Pers,"Weed Research, vol. 34, no. 4, pp. 283–288, 1994.
33. A. S. Collins, C. A. Chase, W. M. Stall, and C. M. Hutchinson, "Competitiveness of three leguminous cover crops with yellow nutsedge (Cyperus esculentus) and smooth pigweed (Amaranthus hybridus),"Weed Science, vol. 55, no. 6, pp. 613–618, 2007.
34. A. Hector, "Overyielding and stable species coexistence," New Phytologist, vol. 172, no. 1, pp. 1–3, 2006.
35. S. R. Radosevich, J. S. Holt, and C. M. Ghersa, Eclogy of Weeds and Invasive Plants: Relatioship to Agriculture and Natural Resource Management, John Wiley & Sons, Hoboken, NJ, USA, 2007.

36. J. A. Mosjidis and G. Wehtje, "Weed control in sunn hemp and its ability to suppress weed growth,"Crop Protection, vol. 30, no. 1, pp. 70–73, 2011.
37. D. R. Gealy, E. J. Wailes, L. E. Estorninos Jr., and R. S. C. Chavez, "Rice cultivar differences in suppression of barnyardgrass (Echinochloa crus-galli) and economics of reduced propanil rates," Weed Science, vol. 51, no. 4, pp. 601–609, 2003.
38. J. S. Holt and D. R. Orcutt, "Functional relationships of growth and competitiveness in perennial weeds and cotton (Gossypium hirsutum)," Weed Science, vol. 39, pp. 575–584, 1991.
39. C. Manidool, "Axonopus compressus (Swartz) P. Beauv," in Plant Resources of South East Asia. 4. Forages, L. Mannetje and R. M. Jones, Eds., Prosea, Bogor, Indonesia, 1992.
40. K. B. Quah, J. K. Lim, and J. Badrulisan, "Alternating herbicide to minimize weed succession in the plantation, in," in Proceedings of the 2nd International Pesticides Conference: Crop Protection towards 2000, Kuala Lumpur, Malaysia, 1997.
41. T. Subtropen, Spatial Undergrowth Species Composition in Oil Palm Plantation in West Sumatra, Palm Oil Institute of Malaysia (PORIM), Kuala Lumpur, Malaysia, 2003.
42. C. W. S. Hartley, The Oil Palm, Longman, London, UK, 3rd edition, 1988.

Chapter 2

Holistic Management: Misinformation on the Science of Grazeded Ecosystems

John Carter, [1] Allison Jones, [2] Mary O'Brien, [3] Jonathan Ratner, [4], and George Wuerthner[5]

[1]Kiesha's Preserve, Paris, ID 83261, USA

[2]Wild Utah Project, Salt Lake City, UT 84101, USA

[3]Grand Canyon Trust, Flagstaff, AZ 86001, USA

[4]Western Watersheds Project, Pinedale, WY 82941, USA

[5]Foundation for Deep Ecology, Bend, OR 97708, USA

ABSTRACT

Over 3 billion hectares of lands worldwide are grazed by livestock, with a majority suffering degradation in ecological condition. Losses in plant productivity, biodiversity of plant and animal communities, and carbon storage are occurring as a result of livestock grazing. Holistic management (HM) has been proposed as a means of restoring degraded deserts and grasslands and reversing climate change. The fundamental approach of this system is based on frequently rotating livestock herds to mimic native ungulates reacting to predators in order to break up biological soil crusts and trample plants and soils to promote restoration. This review could find no peer-reviewed studies that show that this management approach is superior to conventional grazing systems in outcomes. Any claims of success due to HM are likely due to the management aspects of goal setting, monitoring, and adapting to meet goals, not the ecological principles embodied in HM. Ecologically, the application of HM principles of trampling and intensive foraging are as detrimental to plants, soils, water storage, and plant productivity as are conventional grazing systems. Contrary to claims made that HM will reverse climate change, the scientific evidence is that global greenhouse gas emissions are vastly larger than the capacity of worldwide grasslands and deserts to store the carbon emitted each year.

INTRODUCTION

Lands grazed by livestock include 3.4 billion ha worldwide with 73% estimated to be suffering soil degradation [1]. The solution presented during Allan Savory's February 2013 TED Talk was to use holistic management (HM) to reverse desertification and climate change [2]. He reported that we are creating "too much bare ground" (1:30 in video) in the arid areas of the world and, as a consequence, rainfall runs off or evaporates, soils are damaged, and carbon is released back to the atmosphere. Grasslands, even in high rainfall areas, may contain large areas of bare ground with a crust of algae, leading to increased runoff and evaporation. Desertification is caused by livestock, "overgrazing the plants, leaving the soil bare, and giving off methane" (4:20 in video).

HM is also called holistic resource management, time controlled grazing, Savory grazing method, or short-duration grazing. It is designed to mimic the behavior of grazing animals that are regulated by their predators to gather in large groups. As Savory puts it [2], "What we had failed to understand was that these seasonal humidity environments of the world, the soil and the vegetation developed with very large numbers of grazing animals, and that these grazing animals developed with ferocious pack-hunting predators. Now, the main defense against pack-hunting predators is to get into herds, and the larger the herd, the safer the individuals. Now, large herds dung and urinate all over their own food, and they have to keep moving, and it was that movement that prevented the overgrazing of plants, while the periodic trampling ensured good cover of the soil, as we see where a herd has passed." (9:28 in video). In view of the large amount of attention received from the TED talk it is important to examine the validity of Savory's claims.

Savory's writings lack specifics that could be used for implementation of HM or for scientific testing. Details regarding setting of stocking rates, allowable use by livestock, amount of rest needed for recovery, or ecological criteria to be met for biodiversity, sustainability, wildlife, and watershed protection are absent [3–7]. These publications by Savory and his colleagues show that HM is based on the following assumptions: (1) plant communities and soils of the arid, semiarid, and grassland systems of the world evolved in the presence of large herds of animals regulated by their predators; (2) grasses in these areas will become decadent and die out if not grazed by these large herds or their modern day equivalent, livestock; (3) rest from grazing by these large herds of livestock will result in grassland deterioration; (4) large herds are needed to break up decadent plant material and soil crusts and trample dung, urine, seeds, and plant material into the soil, promoting plant growth; and (5) high intensity grazing of these lands by livestock will reverse desertification and climate change by increasing production and cover of the soil, thereby storing more carbon.

We address these five assumptions of HM with a focus on western North American arid and semiarid ecosystems, principally in the desert, steppe, grassland, and open conifer woodland biomes as described by [8]. We use the broad term, grassland, to be inclusive of these types.

ARE WESTERN NORTH AMERICAN ECOSYSTEMS ADAPTED TO HERDS OF LARGE HOOVED ANIMALS?

Not all of today's grasslands, arid, and semiarid systems evolved with herds of large, hooved animals. The Great Plains of North America and subtropical grasslands in Africa that receive moisture during the long, warm, and moist growing season historically supported millions of herbivores [9–11]. Lands west of the Continental Divide of the USA, including the Great Basin, Sonoran, Mojave, and Colorado Plateau deserts, along with the Palouse Prairie grasslands of eastern Washington, western Montana, and northern Idaho, did not evolve with significant grazing pressure from bison (Bison bison) [9, 12, 13]. Though bison were abundant east of the Rockies on the Great Plains, they only occurred in limited numbers across western Wyoming, northeastern Utah, and southeastern Idaho [12, 14]. These low numbers and patchy occurrence would not have played the same ecological role as in the plains. Historically, pronghorn antelope (Antilocapra americana) were more widespread than bison west of the Rockies, but these animals are smaller and lighter than bison and are not ecologically comparable [9]. Evidence for this general lack of large herds of grazing animals west of the Rockies also includes the lack of native dung beetles in the region. Whereas 34 native species of dung beetle (g. Onthophagus) are found east of the Rockies on the plains where bison were numerous, none are found west of the Rockies [15].

The supposition that current western North American plant communities are adapted to livestock grazing because the region supported a diverse herbivore fauna during the Pleistocene epoch ignores that the plant communities have changed in the intervening time [16]. There was rapid evolutionary change following the Pleistocene glaciations in North America with "the establishment of open xeric grasslands in the west central part of the USA... less than 10,000 years ago" [17]. Many of the grasses of the Pleistocene have disappeared from the plains and western USA and the fauna of the Pleistocene was altered by the arrival of bison from Eurasia. These were destructive to long-leaved bunchgrasses found west of the Rockies, while the rhizomatous grasses found east of the Rockies in the prairies were more resistant to their grazing pressure. These rhizomatous grasses are the types found in the prairies of the central and western USA in conjunction with fossil remains of bison. In summary, the western USA of the Pleistocene is not the western USA of today. The climate was much wetter and cooler and the vegetation more mesic in the Pleistocene than today [8, 18]. The drier periods following the Pleistocene as temperatures warmed have altered soil conditions and fire cycles and contributed to the changing flora [19, 20].

Grasslands cover large areas that could support forests but are maintained by grasses outcompeting woody species for belowground resources and providing fuel for fires that limit encroachment of woody species [21]. Climate change and increasing CO_2concentrations have been implicated in the post industrial expansion of woody species and invasive species into grasslands; however, studies from paired locations in grasslands have shown that under similar climatic and CO_2 environments, herbivory by domestic livestock has caused the shifts in woody species and increased invasive species [20, 22, 23]. Elevated CO_2 and climate warming appear to contribute to, but do not explain, the shrub encroachment in these semiarid areas which is due to intensive grazing by domestic livestock [24].

Conclusion. Western US ecosystems outside the prairies in which bison occurred are not adapted to the impact of large herds of livestock. Recent changes to these grassland ecosystems result from herbivory by domestic livestock which has altered fire cycles and promoted invasive species at the expense of native vegetation.

DO GRASSES SENESCE AND DIE IF NOT GRAZED BY LIVESTOCK?

A major premise of HM is that grass species depend on large grazing ungulates in some way, and thus grasses become moribund and die if not grazed, leading to deterioration or eventual loss of the entire grass community [6]. The dead or dormant residual leaves and stalks that remain attached to ungrazed grasses at the end of each growing season can be deceptive. The plants are still alive and healthy, with living buds at the plant base. Bunchgrass canopies collect snow and funnel rainwater to the plant base and soil and increase infiltration [25]. The plants and dead leaves in contact with the soil reduce overland flow and erosion [26]. Plant canopies moderate temperature and protect the growing points from temperature extremes [27]. The standing dead litter provides cover and food for wildlife species including large and small mammals, ground-living birds, and insects. The loss of these leaves and stems through heavy grazing by livestock which occurs under HM destroys these natural attributes.

Grasses with attached dead leaves are more productive than grasses from which the dead leaves have been removed. Loss of these dead tissues to grazers increases thermal damage to the growing shoots and reduces the vigor of the entire plant [28]. Dead leaves and flowering stalks on ungrazed grasses inhibit livestock grazing, allowing those grasses to grow larger than their neighbors [29]. Grazing and trampling by domestic livestock damage plants in natural plant communities [30–32], reduce forage production as stocking rates increase [33], and can lead to simplification of plant communities, establishment of woody vegetation in grasslands, and regression to earlier successional stages [20] or conversion to invasive dominated communities [23] and altered fire cycles [20]. In contrast to the assertion that grasses will die if not grazed by livestock, bunchgrasses in arid environments are more likely to die if they are heavily grazed by domestic animals [34, 35].

Conclusion. Grasses, particularly bunchgrasses, have structure that protects growing points from damage, harvests water, and protects the soil at the plant base. Removal of the standing plant material exposes the growing points, leading to loss or replacement by grazing tolerant species, including invasives.

DOES REST CAUSE GRASSLAND DETERIORATION?

Another principle of HM is that grasslands and their soils deteriorate from overrest, a term that implies insufficient grazing by livestock. However, grasslands that have never been grazed by livestock have been found to support high cover of grasses and forbs. Relict sites throughout the western USA, such as on mesa tops, steep gorges, cliff sides, and even highway rights of way, which are inaccessible to livestock or most ungulates, can retain thriving bunchgrass communities [36–38]. For example, herbaceous growth was vigorous on never-grazed Jordan Valley kipukas in southeast Oregon [37] and on a once-grazed butte called The Island in south central Oregon [36]. Published comparisons of grazed and ungrazed lands in the western USA have found that rested sites have larger and more dense grasses, fewer weedy forbs and shrubs, higher biodiversity, higher productivity, less bare ground, and better water infiltration than nearby grazed sites. These reports include 139 sites in south Dakota [39], as well as sites that had been rested for 18 years in Montana [40], 30 years in Nevada [41], 20–40 years in British Columbia [42], 45 years in Idaho [43], and 50 years in the Sonoran Desert of Arizona [44]. None of the above studies demonstrated that long periods of rest damaged native grasslands. A list and description of such sites can be found in [45].

The HM misinterpretation of the natural history of grazed and ungrazed grasslands is apparent in Savory's description of the Appleton-Whittell Research Ranch in southeastern Arizona [6]. This ranch has been protected from livestock grazing since 1968 with the grasses on the ranch described by Savory as becoming "moribund" (page 211), with "bare spots opening up" (page 211). In contrast to those claims, plant species richness on the ranch increased from 22 species in 1969 to 49 species in 1984, while plant cover increased from 29% in 1968 to 85% in 1984 [46]. Total grass cover on the ranch was significantly higher on ungrazed sites when compared to grazed sites ($P<0.01$) [47]. These well designed studies produced quantitative data showing that the HM view of the ranch is not the case.

Conclusion. Contrary to the assumption that grasses will senesce and die if not grazed by livestock, studies of numerous relict sites, long-term rested sites, and paired grazed and ungrazed sites have demonstrated that native plant communities, particularly bunchgrasses, are sustained by rest from livestock grazing.

IS HOOF ACTION NECESSARY FOR GRASSLAND HEALTH?

A key premise of HM is that livestock can be made to emulate native ungulate responses to predators by moving them frequently in large numbers and tight groups. This promotes very close cropping which is said to benefit grasses and other forage, as well as hoof action that breaks up soil crusts, increases infiltration, plants seeds, and incorporates plant material, manure, and urine into the soil [6]. Other than bison in the plains states, the evidence indicates a low frequency of large hooved mammals in the western USA during pre-Columbian times [48], so the opportunity for hoof action to sustain grasslands and deserts appears limited at best. In contrast to HM claims, elk (Cervus canadensis), mule deer (Odocoileus hemionus), and other ungulates may avoid areas where predators have an advantage in capturing them [49]. Avoidance is not the same as a panicked flight or tight groupings of animals promoting hoof action. Rather, the major response is greater vigilance and sometimes avoidance of risky areas. While the presence of wolves (Canis lupus) affects elk behavior by reducing browsing on willows and aspen [50], snow depth and other ecological needs appear to outweigh the effect of wolves leading to grazing and browsing in areas of higher risk [51, 52]. We found no documentation of native animal responses to predators generating hoof action or herd effect in tight groupings in the western USA.

Soils in arid and semiarid grasslands often have significant areas covered by biological crusts [53–55]. These are made up of bacteria, cyanobacteria, algae, mosses, and lichens and are essential to the health of these grasslands. Biological crusts stabilize soils, increase soil organic matter and nutrient content, absorb dew during dry periods, and fix nitrogen [53, 56–60]. Crusts enhance soil stability and reduce water runoff by producing more microcatchments on soil surfaces. They increase water absorbing organic matter, improve nutrient flow, germination and establishment for some plants, while dark crusts may stimulate plant growth by producing warmer soil temperatures and water uptake in cold deserts [61]. Some crusts are hydrophobic, shedding water [60]. Biological soil crusts are fragile, highly susceptible to trampling [61–63], and are slow to recover from trampling impacts [64]. Loss of these crusts results in increased erosion and reduced soil fertility. The loss of crusts in the bunchgrass communities of the western USA may be largely responsible for the widespread establishment of cheatgrass and other exotic annuals [23,58, 65]. The rapid spread of introduced weeds throughout the arid western USA is estimated at over 2000 hectares per day [66], largely due to livestock disturbance.

The HM assumption that increasing hoof action will increase infiltration has been disproven. Livestock grazing can compact soil, reduce infiltration, and increase runoff, erosion, and sediment yield [67–71]. Major increases in erosion and runoff occur under normal stocking when comparing grazed to ungrazed sites [68, 71–74]. Extensive literature reviews report the negative impacts of livestock grazing on soil stability and erosion [75–77]. For example, a study of wet and dry meadows in Oregon found the infiltration rate in ungrazed dry meadows was 13 times greater and 2.3 times greater in ungrazed wet meadows, compared to similar grazed meadows [78].

Hoof action is not needed to increase soil fertility and decomposition of litter. It is well-established that soil protozoa, arthropods, earthworms, microscopic bacteria, and fungi decompose plant and animal residues in all environments [79, 80]. Even the driest environments contain 100 million to one billion decomposing bacteria and tens to hundreds of meters of fungal hyphae per gram of soil [81]. Brady and Weil [80] discuss the importance of mammals in the decay process, mentioning burrowing mammals, but not large grazers such as cattle and bison. Removal of plant biomass and lowered production resulting from livestock grazing can reduce fertility and organic content of the soil [70,82–84].

Conclusion. We found no evidence that hoof action as described by Savory occurs in the arid and semiarid grasslands of the western USA which lacked large herds of ungulates such as bison that occurred in the prairies of the USA or the savannahs of Africa. No benefits of hoof action were found. To the contrary, hoof action by livestock has been documented to destroy biological crusts, a key component in soil protection and nutrient cycling, thereby increasing erosion rates and reducing fertility, while, increasing soil compaction and reducing water infiltration.

CAN GRAZING LIVESTOCK INCREASE CARBON STORAGE AND REVERSE CLIMATE CHANGE?

Among the most recent HM claims is that livestock grazing will lead to sequestration of large amounts of carbon, thus potentially reversing climate change [2]. However, any increased carbon storage through livestock grazing must be weighed against the contribution of livestock metabolism to greenhouse gas emissions due to rumen bacteria methane emissions, manure, and fossil fuel use across the production chain [85, 86]. Nitrous oxide, 300 times more potent than methane in trapping greenhouse gases [87], is also produced and released with livestock production. The livestock industry's contribution to greenhouse gases also includes CO_2 released by conversion of forests to grasslands for the purpose of grazing [86].

Worldwide, livestock production accounts for about 37 percent of global anthropogenic methane emissions and 65 percent of anthropogenic nitrous oxide emissions with as much as 18% of current global greenhouse gas emissions (CO_2 equivalent) generated from the livestock industry [85]. It is estimated that livestock production, byproducts, and other externalities account for 29.5 billion metric tons of CO_2 per year or 51 percent of annual worldwide greenhouse gas emissions from agriculture [88]. Lower amounts of greenhouse gas emissions due to livestock may be estimated by using narrower definitions of livestock-related emissions that include feed based emissions only and exclude externalities [89].

Some suggest that grass-fed beef is a superior alternative to beef produced in confined animal feeding operations [90]. However, grass provides less caloric energy per pound of feed than grain and, as a consequence, a grass-fed cow's rumen bacteria must work longer breaking down and digesting grass in order to extract the same energy content found in grain, while the bacteria in its rumen are emitting methane [89]. Comparisons of pasture-finished and feedlot-finished beef in the USA found that pasture-finished beef produced 30% more greenhouse gas emissions on a live weight basis [91].

It is estimated that three times as much carbon resides in soil organic matter as in the atmosphere [92], while grasslands and shrublands have been estimated to store 30 percent of the world's soil carbon with additional amounts stored in the associated vegetation [93]. Long term intensive agriculture can significantly deplete soil organic carbon [94] and past livestock grazing in the United States has led to such losses [95, 96]. Livestock grazing was also found to significantly reduce carbon storage on Australian grazed lands while destocking currently grazed shrublands resulted in net carbon storage [97]. Livestock-grazed sites in Canyonlands National Park, Utah, had 20% less plant cover and 100%

Conclusion. HM does not address riparian areas and biodiversity with its focus on livestock production, although operators could choose these as goals. We have seen no studies of HM impacts on riparian areas and biodiversity, although livestock grazing impacts on riparian areas and biodiversity have been well documented. Livestock degrade riparian areas by removal of streamside vegetation, reduction of cover and food for fish and wildlife, and soil compaction, erosion, and sedimentation. These impacts lead to loss of native fish and wildlife populations. Studies in areas from which livestock have been removed demonstrate increases in diversity and abundance of birds, mammals, insects, and fish.

IS SCIENTIFIC EVIDENCE IMPORTANT?

Effectiveness studies of HM have been undertaken by ranchers and farmers who were selected because of their commitment to HM [119]. In other words, such studies were neither experimental nor were the participants randomly selected. Livestock producers who may have had negative experiences with HM were not included in the studies. Nearly all of the support and confirmation for HM come from articles developed at the Savory Institute or testimonials by practitioners. Most of the published literature that attempts to rigorously test HM in any scientific fashion does not support its principal assumptions.

Holechek et al. [83] stated that "No grazing approach, including that of Savory, will overcome the adverse effects of drought and/or chronic heavy stocking on forage production." These researchers were also critical of government agencies for adopting these unproven theories rather than basing management on "scientifically proven range management practices and principles" [83] (page 25). Briske et al. [103] stated,

The rangeland profession has become mired in confusion, misinterpretation, and uncertainty with respect to the evaluation of grazing systems and the development of grazing recommendations and policy decisions. We contend that this has occurred because recommendations have traditionally been based on perception, personal experience, and anecdotal interpretations of management practices, rather than evidence-based assessments of ecosystem responses. [103] (page 11).

Briske et al. [102] state, "Mr. Savory's attempts to divide science and management perspectives and his aggressive promotion of a narrowly focused and widely challenged grazing method only serve to weaken global efforts to promote rangeland restoration and C sequestration." [102] (page 74).

Conclusion. Studies supporting HM have generally come from the Savory Institute or anecdotal accounts of HM practitioners. Leading range scientists have refuted the system and indicated that its adoption by land management agencies is based on these anecdotes and unproven principles rather than scientific evidence. When addressing the application of HM or any other grazing systems, practitioners, including agencies managing public lands, private livestock operations, and scientists, should (1) consider inclusion of watershed-scale ungrazed reference areas of suitable size to encompass the plant and soil communities found in the grazed area, (2) define ecological (plant, soil, and animal community) and production (livestock) criteria on which to base quantitative comparisons, (3) use sufficient replication in studies, (4) and include adequate quality control of methods. Economic analysis of grazing systems should compare all expenditures with income, including externalized costs such as soil loss, water pollution, reduction of water infiltration, and carbon emissions and capture.

MANAGEMENT IMPLICATIONS

This review shows that the underlying assumptions of HM regarding the evolutionary adaptation of western North American landscapes to large herds of hooved animals only applies to prairie grasslands and that most arid and semiarid areas of western North America are not adapted to their impacts. The premise that rest results in degradation of grassland ecosystems by allowing biological crusts to persist and grasses to senesce and die has been disproven by a large body of research. Reliance on hoof action to promote recovery by trampling seeds and organic matter into the soil and breaking up soil crusts needs to be considered in the context of increased soil compaction, lower infiltration rates, and the destruction of biological crusts that normally provide long-term stability to soil surfaces, enhance water retention, and promote nutrient cycling. The use of HM in an attempt to capture atmospheric greenhouse gases and incorporate them into soils and plant communities, thereby reducing climate change effects, is demonstrably impossible because the nonforested grazed lands of the world do not have the capacity to sequester this amount of emissions. Even in the prairie regions of the United States, which are evolutionarily adapted to large herbivores such as bison, research indicates that not only does HM not produce results superior to conventional season-long grazing, but also that stocking rate, rest, and livestock exclusion represent the best mechanisms for restoring grassland productivity, ecological condition, and sustainability. Various studies indicate livestock grazing reduces biodiversity of native species and degrades riparian areas, with nearly all studies finding livestock exclusion to be the most effective, reliable means to restore degraded riparian areas. Claims of the benefits of HM or other grazing systems should be validated by quantitative, scientifically valid studies.

ACKNOWLEDGMENT

Joy Belsky (1944–2001), Range Ecologist, who made available a draft analysis of holistic management before her death, provided us with much of the material presented here.

REFERENCES

1. E. Gabathuler, H. Liniger, C. Hauert, and M. Giger, Benefits of Sustainable Land Management, World Overview of Conservation Approaches and Technologies, Center for Development and Environment, University of Bern, Bern, Switzerland, 2009.
2. A. Savory, "How to fight desertification and reverse climate change," 2013,http://www.ted.com/talks/allan_savory_how_to_green_the_world_s_deserts_and_reverse_climate_change.html.
3. A. Savory and S. D. Parsons, "The Savory grazing method," Rangelands, vol. 2, pp. 234–237, 1980.
4. A. Savory, "The Savory grazing method or holistic resource management,"Rangelands, vol. 5, pp. 155–159, 1983.

5. A. Savory, Holistic Resource Management, Island Press, Washington, DC, USA, 1988.
6. A. Savory and J. Butterfield, Holistic Management: A New Framework for Decision Making, Island Press, Washington, DC, USA, 1999.
7. C. J. Hadley, "The wild life of Allan Savory," Rangelands, vol. 22, pp. 6–10, 2000.
8. R. S. Thompson and K. H. Anderson, "Biomes of western North America at 18,000, 6000 and 0 ^{14}C yr BP reconstructed from pollen and packrat midden data," Journal of Biogeography, vol. 27, no. 3, pp. 555–584, 2000.
9. R. N. Mack and J. N. Thompson, "Evolution in steppe with few large hooved mammals," American Naturalist, vol. 119, no. 6, pp. 757–773, 1982.
10. A. R. E. Sinclair and M. Norton-Griffiths, Serengeti: Dynamics of an Ecosystem, University of Chicago Press, 1979.
11. A. R. E. Sinclair, S. A. R. Mduma, J. G. C. Hopcraft, J. M. Fryxell, R. Hilborn, and S. Thirgood, "Long-term ecosystem dynamics in the serengeti: lessons for conservation," Conservation Biology, vol. 21, no. 3, pp. 580–590, 2007.
12. R. Daubenmire, "The western limits of the range of the American bison," Ecology, vol. 66, no. 2, pp. 622–624, 1985.
13. G. Wuerthner, "Are cows just domestic bison? Behavioral and habitat use differences between cattle and bison," in Proceedings of an International Symposium on Bison Ecology and Management in North America, L. Irby, L. Knight, and J. Knight, Eds., pp. 374–383, Bozeman, Mont, USA, June 1998.
14. F. G. Roe, The North American Buffalo: A Critical Study of the Species in Its Wild State, University of Toronto Press, Toronto, Calif, USA, 1951.
15. J. F. Howden, "Some possible effects of the Pleistocene on the distributions of North American Scarabaeidae (Coleoptera)," Canadian Entomologist, vol. 98, no. 11, pp. 1177–1190, 1966.
16. J. W. Burkhardt, Herbivory in the Intermountain West, vol. 58 of Station Bulletin, University of Idaho Forest, Wildlife and Range Experiment Station, Moscow, Idaho, USA, 1996.

17. J. M. J. de Wet, "Grasses and the culture history of man," Annals Missouri Botanical Garden, vol. 68, no. 1, pp. 87–104, 1981.
18. H. Wanner, J. Beer, J. Bütikofer et al., "Mid- to Late Holocene climate change: an overview," Quaternary Science Reviews, vol. 27, no. 19-20, pp. 1791–1828, 2008.
19. D. K. Grayson, "Mammalian responses to middle Holocene climatic change in the Great Basin of the western United States," Journal of Biogeography, vol. 27, no. 1, pp. 181–192, 2000.
20. R. L. Beschta, D. L. Donahue, A. DellaSala et al., "Adapting to climate change on western public lands: addressing the ecological effects of domestic, wild, and feral ungulates," Environmental Management, vol. 51, no. 2, pp. 474–491, 2012.
21. W. J. Bond, "What limits trees in C4 grasslands and savannas?" Annual Review of Ecology, Evolution, and Systematics, vol. 39, pp. 641–659, 2008.
22. S. Archer, D. S. Schimel, and E. A. Holland, "Mechanisms of shrubland expansion: land use, climate or CO_2?" Climatic Change, vol. 29, no. 1, pp. 91–99, 1995.
23. M. D. Reisner, J. B. Grace, D. A. Pyke, and P. S. Doescher, "Conditions favoringBromus tectorum dominance of endangered sagebrush steppe ecosystems," Journal of Applied Ecology, vol. 50, no. 4, pp. 1039–1049, 2013.
24. O. W. van Auken, "Shrub invasions of North American semiarid grasslands,"Annual Review of Ecology and Systematics, vol. 31, pp. 197–215, 2000.
25. G. W. Gee, P. A. Beedlow, and R. L. Skaggs, "Water balance," in Shrub-Steppe Balance and Change in a Semi-Arid Terrestrial Ecosystem, W. H. Rickard, L. E. Rogers, B. E. Vaughan, and S. F. Liebetrau, Eds., pp. 61–81, Elsevier, New York, NY, USA, 1988.
26. W. H. Wischmeier and D. D. Smith, Predicting Rainfall Erosion Losses: A Guide to Conservation Planning, vol. 537 of Agriculture Handbook, US Department of Agriculture, Washington, DC, USA, 1978.
27. W. T. Hinds and W. H. Rickard, "Soil temperatures near a desert steppe shrub,"Northwest Science, vol. 42, pp. 5–8, 1968.

28. R. H. Sauer, "Effect of removal of standing dead material on growth of Agropyron spicatum," Journal of Range Management, vol. 31, no. 2, pp. 121–122, 1978.
29. D. Ganskopp, R. Angell, and J. Rose, "Response of cattle to cured reproductive stems in a caespitose grass," Journal of Range Management, vol. 45, no. 4, pp. 401–404, 1992.
30. L. Ellison, "Influence of grazing on plant succession of Rangelands," The Botanical Review, vol. 26, no. 1, pp. 1–78, 1960.
31. A. J. Belsky, "Does herbivory benefit plants? A review of the evidence," American Naturalist, vol. 127, no. 6, pp. 870–892, 1986.
32. E. L. Painter and A. J. Belsky, "Application of herbivore optimization theory to rangelands of the western United States," Ecological Appplications, vol. 3, no. 1, pp. 2–9, 1993.
33. J. L. Holechek, H. Gomez, F. Molinar, and D. Galt, "Grazing studies: what we've learned," Rangelands, vol. 21, no. 2, pp. 12–16, 1999.
34. W. F. Mueggler, "Rate and pattern of vigor recovery in Idaho fescue and bluebunch wheatgrass," Journal of Range Management, vol. 28, no. 3, pp. 198–204, 1975.
35. L. D. Anderson, Bluebunch Wheatgrass Defoliation, Effects and Recovery—A Review, vol. 91-2 of BLM Technical Bulletin, Bureau of Land Management, Idaho State Office, Boise, Idaho, USA, 1991.
36. R. S. Driscoll, "A relict area in the central Oregon juniper zone," Ecology, vol. 45, no. 2, pp. 345–353, 1964.
37. R. R. Kindschy, "Pristine vegetation of the Jordan Crater kipukas: 1978–1991," inProceedings-Ecology and Management of Annual Rangelands, S. B. Monsen and S. G. Kitchen, Eds., INT-GTR-313, pp. 85–88, US Department of Agriculture, Forest Service, Boise, Idaho, USA, May 1992.
38. N. Ambos, G. Robertson, and J. Douglas, "Dutchwoman butte: a relict grassland in central Arizona," Rangelands, vol. 22, no. 2, pp. 3–8, 2000.
39. D. F. Costello and G. T. Turner, "Vegetation changes following exclusion of livestock from grazed ranges," Journal of Forestry, vol. 39, pp. 310–315, 1941.

40. A. B. Evanko and R. A. Peterson, "Comparisons of protected and grazed mountain rangelands in southwestern Montana," Ecology, vol. 36, no. 1, pp. 71–82, 1955.
41. J. H. Robertson, "Changes on a sagebrush-grass range in Nevada ungrazed for 30 years," Journal of Range Management, vol. 24, no. 5, pp. 397–400, 1971.
42. A. McLean and E. W. Tisdale, "Recovery rate of depleted range sites under protection from grazing," Journal of Range Management, vol. 25, no. 3, pp. 178–184, 1972.
43. J. E. Anderson and R. S. Inouye, "Landscape-scale changes in plant species abundance and biodiversity of a sagebrush steppe over 45 years," Ecological Monographs, vol. 71, no. 4, pp. 531–556, 2001.
44. J. Blydenstein, C. R. Hungerford, G. I. Day, and R. R. Humphrey, "Effect of domestic livestock exclusion on vegetation in the Sonoran Desert," Ecology, vol. 38, no. 3, pp. 522–526, 1957.
45. G. Wuerthner and M. Matteson, "A guide to livestock-free landscapes," in Welfare Ranching: The Subsidized Destruction of the American West, G. Wuerthner and M. Mattson, Eds., pp. 327–329, Island Press, Washington, DC, USA, 2004.
46. W. W. Brady, M. R. Stromberg, E. F. Aldon, C. D. Bonham, and S. H. Henry, "Response of a semidesert grassland to 16 years of rest from grazing," Journal of Range Management, vol. 42, no. 4, pp. 284–288, 1989.
47. C. E. Bock, J. H. Bock, W. R. Penney, and V. M. Hawthorne, "Responses of birds, rodents, and vegetation to livestock exclosure in a semidesert grassland site," Journal of Range Management, vol. 37, no. 3, pp. 239–242, 1984.
48. W. J. Ripple and B. van Valkenburgh, "Linking top-down forces to the pleistocene megafaunal extinctions," BioScience, vol. 60, no. 7, pp. 516–526, 2010.
49. J. W. Laundré, L. Hernández, and W. J. Ripple, "The landscape of fear: ecological implications of being afraid," The Open Ecology Journal, vol. 3, no. 2, pp. 1–7, 2010.
50. C. Eisenberg, S. T. Seager, and D. E. Hibbs, "Wolf, elk, and aspen food web relationships: context and complexity," Forest Ecology and Management, vol. 299, pp. 70–80, 2013.

51. S. Creel and D. Christianson, "Wolf presence and increased willow consumption by Yellowstone elk: implications for trophic cascades," Ecology, vol. 90, no. 9, pp. 2454–2466, 2009.
52. J. Winnie Jr. and S. Creel, "Sex-specific behavioural responses of elk to spatial and temporal variation in the threat of wolf predation," Animal Behaviour, vol. 73, no. 1, pp. 215–225, 2007.
53. J. Belnap, D. Eldridge, J. H. Kaltenecker, S. Leonard, R. Rosentreter, and J. Williams,Biological Soil Crusts Ecology and Management, TR-1730-2, US Department of Interior, Bureau of Land Management, Denver, Colo, USA, 2001.
54. N. E. West, "Western intermountain sagebrush steppe," in Temperate Deserts and Semi-Deserts, N. E. West, Ed., pp. 351–373, Elsevier Scientific Publishing Company, Amsterdam, The Netherlands, 1983.
55. J. Belnap and O. L. Lange, Eds., Biological Soil Crusts: Structure, Function, and Management, Springer, New York, NY, USA, 2003.
56. O. L. Lange, E. D. Schulze, L. Kappen, U. Buschbom, and M. Evenari, "Adaptations of desert lichens to drought and extreme temperatures," in Environmental Physiology of Desert Ecosystems, N. F. Hadley, Ed., pp. 27–30, Dowden, Hutchinson and Ross, Stroudsberg, Pa, USA, 1975.
57. J. A. R. Ladyman and E. Muldavin, Terrestrial Cryptogams of Pinyon-Juniper Woodlands in the Southwestern US: A Review, RM-GTR-280, US Department of Agriculture, Forest Service, Fort Collins, Colo, USA, 1996.
58. A. J. Belsky and J. L. Gelbard, Livestock Grazing and Weed Invasions in the Arid West, Oregon Natural Desert Association, Bend, Ore, USA, 2000.
59. G. Wuerthner, "The soil's living surface: biological crusts," in Welfare Ranching: The Subsidized Destruction of the American West, G. Wuerthner and M. Mattson, Eds., pp. 199–204, Island Press, Washington, DC, USA, 2004.
60. L. Deines, R. Rosentreter, D. J. Eldridge, and M. D. Serpe, "Germination and seedling establishment of two annual grasses on lichen-dominated biological soil crusts," Plant and Soil, vol. 295, no. 1-2, pp. 23–35, 2007.
61. J. Belnap, "Potential role of cryptobiotic soil crust in semi-arid rangelands," inProceedings-Ecology and Management of Annual

105. J. M. Earl and C. E. Jones, "The need for a new approach to grazing management—is cell grazing the answer?" The Rangeland Journal, vol. 18, no. 2, pp. 327–350, 1996.

106. G. Sanjari, H. Ghadiri, C. A. A. Ciesiolka, and B. Yu, "Comparing the effects of continuous and time-controlled grazing systems on soil characteristics in southeast Queensland," Australian Journal of Soil Research, vol. 46, no. 4, pp. 348–358, 2008.

107. W. R. Teague, S. L. Dowhower, S. A. Baker, N. Haile, P. B. DeLaune, and D. M. Conover, "Grazing management impacts on vegetation, soil biota and soil chemical, physical and hydrological properties in tall grass prairie," Agriculture, Ecosystems and Environment, vol. 141, no. 3-4, pp. 310–322, 2011.

108. K. T. Weber and B. S. Gokhale, "Effect of grazing on soil-water content in semiarid rangelands of southeast Idaho," Journal of Arid Environments, vol. 75, no. 5, pp. 464–470, 2011.

109. H. W. van Poolen and J. R. Lacey, "Herbage response to grazing systems and stocking intensities," Journal of Range Management, vol. 32, no. 4, pp. 250–253, 1979.

110. W. P. Clary and B. F. Webster, "Managing grazing of riparian areas in the Intermountain Region," Tech. Rep. GTR-INT-263, US Department of Agriculture, Forest Service, Ogden, Utah, USA, 1989.

111. H. K. Orr, "Recovery from soil compaction on bluegrass range in the Black Hills,"Transactions of the American Society of Agricultural and Biological Engineers, vol. 18, no. 6, pp. 1076–1081, 1975.

112. A. J. Belsky, A. Matzke, and S. Uselman, "Survey of livestock influences on stream and riparian ecosystems in the western United States," Journal of Soil and Water Conservation, vol. 54, no. 1, pp. 419–431, 1999.

113. T. Tucker Schulz and W. C. Leininger, "Differences in riparian vegetation structure between grazed areas and exclosures," Journal of Range Management, vol. 43, no. 4, pp. 295–299, 1990.

114. E. J. Dyksterhuis, "Condition and management of rangeland based on quantitative ecology," Journal of Range Management, vol. 2, no. 3, pp. 104–115, 1949.

115. E. F. Habich, "Ecological site inventory, technical reference 1734-7," Tech. Rep. BLM/ST/ST-01/003+1734, Bureau of Land Management, Denver, Colo, USA, 2001.
116. D. S. Dobkin, A. C. Rich, and W. H. Pyle, "Habitat and avifaunal recovery from livestock grazing in riparian meadow system of the northwestern Great Basin,"Conservation Biology, vol. 12, no. 1, pp. 209–221, 1998.
117. S. L. Earnst, J. A. Ballard, and D. S. Dobkin, "Riparian songbird abundance a decade after cattle removal on Hart Mountain and Sheldon National Wildlife Refuges," inProceedings of the 3rd International Partners in Flight Conference, C. J. Ralph and T. Rich, Eds., General Technical Report PSW-GTR-191, pp. 550–558, US Department of Agriculture, Forest Service, Albany, Calif, USA, 2005.
118. C. E. Bock and J. H. Bock, "Cover of perennial grasses in southeastern Arizona in relation to livestock grazing," Conservation Biology, vol. 7, no. 2, pp. 371–377, 1993.
119. D. H. Stinner, B. R. Stinner, and E. Martsolf, "Biodiversity as an organizing principle in agroecosystem management: case studies of holistic resource management practitioners in the USA," Agriculture, Ecosystems and Environment, vol. 63, no. 2-3, pp. 199–213, 1997.

Chapter 3

Consequences of Changing Biodiversity

F. Stuart Chapin III[1], Erika S. Zavaleta[2], Valerie T. Eviner[4], Rosamond L. Naylor[3], Peter M. Vitousek[2], Heather L. Reynolds[5], David U. Hooper[6], Sandra Lavorel[7], Osvaldo E. Sala[8], Sarah E. Hobbie[9], Michelle C. Mack[1] & Sandra Díaz[10]

Institute of Arctic Biology, University of Alaska, Fairbanks, Alaska 99775, USA

Department of Biological Sciences and Institute for International Studies, Stanford University, Stanford, California 94305, USA

Department of Integrative Biology, University of California, Berkeley, California 94720, USA

Department of Biology, Kalamazoo College, Kalamazoo, Michigan 49006, USA

¶ Department of Biology, Western Washington University, Bellingham, Washington 98225, USA

Centre d'Ecologie Fonctionnelle et Evolutive, CNRS UPR 9056, 34293 Montpellier Cedex 05, France

Cátedra de Ecología and Instituto de Fisiología y Ecología Vinculadas a la Agricultura, Faculty of Agronomy, University of Buenos Aires, Ave San Martín 4453, Buenos Aires C1417DSE, Argentina

Department of Ecology, Evolution, and Behavior, University of Minnesota, St Paul, Minnesota 55108, USA

Instituto Multidisciplinario de Biología Vegetal, Universidad Nacional de Córdoba, FCEFyN, Casilla de Correo 495, 5000 Córdoba, Argentina

ABSTRACT

Human alteration of the global environment has triggered the sixth major extinction event in the history of life and caused widespread changes in the global distribution of organisms. These changes in biodiversity alter ecosystem processes and change the resilience of ecosystems to environmental change. This has profound consequences for services that humans derive from ecosystems. The large ecological and societal consequences of changing biodiversity should be minimized to preserve options for future solutions to global environmental problems.

INTRODUCTION

Humans have extensively altered the global environment, changing global biogeochemical cycles, transforming land and enhancing the mobility of biota. Fossil-fuel combustion and deforestation have increased the concentration of atmospheric carbon dioxide (CO_2) by 30% in the past three centuries (with more than half of this increase occurring in the past 40 years). We have more than doubled the concentration of methane and increased concentrations of other gases that contribute to climate warming. In the next century these greenhouse gases are likely to cause the most rapid climate change that the Earth has experienced since the end of the last glaciation 18,000 years ago and perhaps a much longer time. Industrial fixation of nitrogen for fertilizer and other human activities has more than doubled the rates of terrestrial fixation of gaseous nitrogen into biologically available forms. Run off of nutrients from agricultural and urban systems have increased several-fold in the developed river basins of the Earth, causing major ecological changes in estuaries and coastal zones. Humans have transformed 40–50% of the ice-free land surface, changing prairies, forests and wetlands into agricultural and urban systems. We dominate (directly or indirectly) about one-third of the net primary productivity on land and harvest fish that use 8% of ocean productivity. We use 54% of the available fresh water, with use projected to increase to 70% by 2050[1]. Finally, the mobility of people has transported organisms across geographical barriers that long kept the biotic regions of the Earth separated, so that many of the ecologically important plant and animal species of many areas have been introduced in historic time[2,3].

Together these changes have altered the biological diversity of the Earth (Fig. 1). Many species have been eliminated from areas dominated by human influences. Even in preserves, native species are often out-competed or consumed by organisms introduced from elsewhere. Extinction is a natural process, but it is occurring at an unnaturally rapid rate as a consequence of human activities. Already we have caused the extinction of 5–20% of the species in many groups of organisms (Fig. 2), and current rates of extinction are estimated to be 100–1,000 times greater than pre-human rates[4,5].

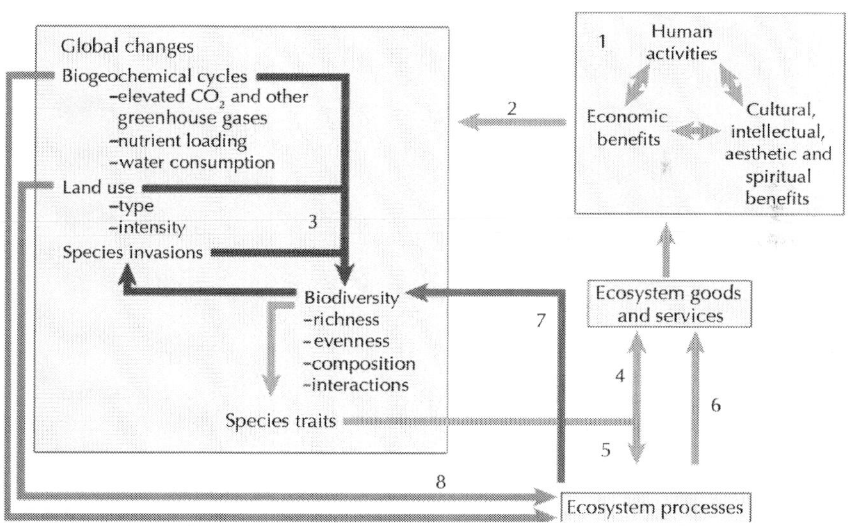

Figure 1: The role of biodiversity in global change. Human activities that are motivated by economic, cultural, intellectual, aesthetic and spiritual goals (1) are now causing environmental and ecological changes of global significance (2). By a variety of mechanisms, these global changes contribute to changing biodiversity, and changing biodiversity feeds back on susceptibility to species invasions (3, purple arrows; see text). Changes in biodiversity, through chang-

es in species traits, can have direct consequences for ecosystem services and, as a result, human economic and social activities (4). In addition, changes in biodiversity can influence ecosystem processes (5). Altered ecosystem processes can thereby influence ecosystem services that benefit humanity (6) and feedback to further alter biodiversity (7, red arrow). Global changes may also directly affect ecosystem processes (8, blue arrows). Depending on the circumstances, the direct effects of global change may be either stronger or weaker than effects mediated by changes in diversity. We argue that the costs of loss of biotic diversity, although traditionally considered to be 'outside the box' of human welfare, must be recognized in our accounting of the costs and benefits of human activities.

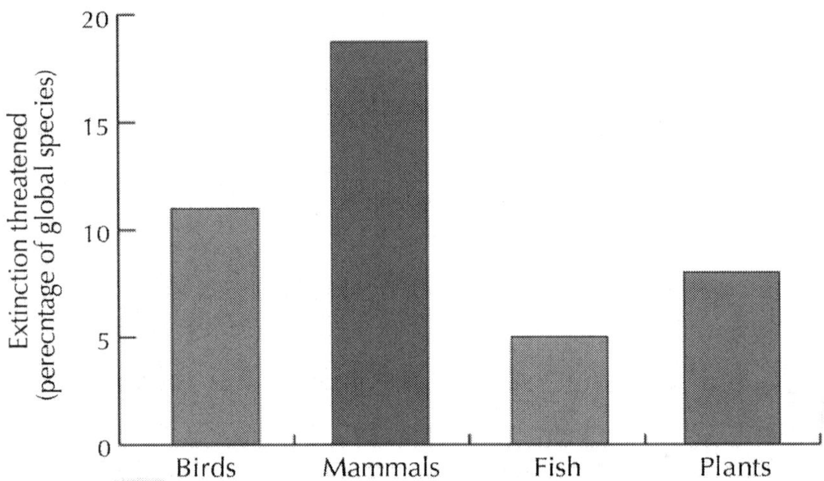

Figure 2: Proportion of the global number of species of birds, mammals, fish and plants that are currently threatened with extinction[4]. In the absence of major changes in policy and human behaviour, our effects on the environment will continue to alter biodiversity. Land-use change is projected to have the largest global impact on biodiversity by the year 2100, followed by climate change, nitrogen deposition, species introductions and changing concentrations of atmospheric CO_2 (ref. 6). Land-use change is expected to be of particular importance in the tropics, climatic change is likely to be important at high latitudes, and a multitude of interacting causes will affect other biomes (Fig. 3)[6]. What are the ecological and societal consequences of current and projected effects of human activity on biological diversity?

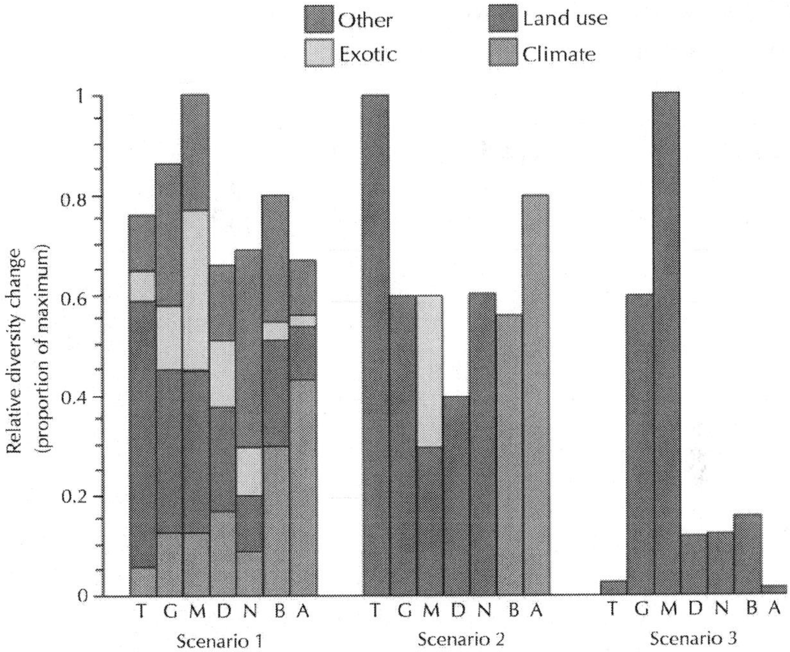

Figure 3: Scenarios of change in species diversity in selected biomes by the year 2100. The values are the projected change in diversity for each biome relative to the biome with greatest projected diversity change[6]. Biomes are: tropical forests (T), grasslands (G), Mediterranean (M), desert (D), north temperate forests (N), boreal forests (B) and arctic (A). Projected change in species diversity is calculated assuming three alternative scenarios of interactions among the causes of diversity change. Scenario 1 assumes no interaction among causes of diversity change, so that the total change in diversity is the sum of the changes caused by each driver of diversity change. Scenario 2 assumes that only the factor with the greatest impact on diversity influences diversity change. Scenario 3 assumes that factors causing change in biodiversity interact multiplicatively to determine diversity change. For scenarios 1 and 2, we show the relative importance of the major causes of projected change in diversity. These causes are climatic change, change in land use, introduction of exotic species, and changes in atmospheric CO_2 and/or nitrogen deposition (labelled 'other'). The graph shows that all biomes are projected to experience substantial change in species diversity by 2100, that the most important causes of diversity change differ among biomes, and that the patterns of diversity change depend on assumptions about the nature of interactions among the causes of diversity change. Projected biodiversity change is most similar among biomes if causes of diversity change do not

interact (scenario 1) and differ most strongly among biomes if the causes of biodiversity change interact multiplicatively (scenario 3).

ECOSYSTEM CONSEQUENCES OF ALTERED DIVERSITY

Diversity at all organizational levels, ranging from genetic diversity within populations to the diversity of ecosystems in landscapes, contributes to global biodiversity. Here we focus on species diversity, because the causes, patterns and consequences of changes in diversity at this level are relatively well documented. Species diversity has functional consequences because the number and kinds of species present determine the organismal traits that influence ecosystem processes. Species traits may mediate energy and material fluxes directly or may alter abiotic conditions (for example, limiting resources, disturbance and climate) that regulate process rates. The components of species diversity that determine this expression of traits include the number of species present (species richness), their relative abundances (species evenness), the particular species present (species composition), the interactions among species (non-additive effects), and the temporal and spatial variation in these properties. In addition to its effects on current functioning of ecosystems, species diversity influences the resilience and resistance of ecosystems to environmental change.

Species Richness and Evenness

Most theoretical and empirical work on the functional consequences of changing biodiversity has focused on the relationship between species richness and ecosystem functioning. Theoretical possibilities include positive linear and asymptotic relationships between richness and rates of ecosystem processes, or the lack of a simple statistical relationship[7] (Box 1). In experiments, species richness correlates with rates of ecosystem processes most clearly at low numbers of species. We know much less about the impact of species richness in species-rich, natural ecosystems. Several studies using experimental species assemblages have shown that annual rates of primary productivity and nutrient retention increase with increasing plant species richness, but saturate at a rather low number of species[8, 9]. Arbuscular mycorrhizal species richness also seems to enhance plant production in an asymptotic fashion, although phosphorus uptake was enhanced in a linear fashion from 1 to 14 species of fungi[10]. Microbial richness can lead to increased decomposition of organic matter[11]. In contrast, no consistent statistical relationship has been observed between plant species richness of litter inputs and decomposition rate[12]. Thus, in experimental communities (which typically focus on only one or two trophic levels), there seems to be no universal relationship between species richness and ecosystem functioning, perhaps because processes differ in their sensitivity to species richness compared with other components of diversity (such as evenness, composition or interactions). The absence of a simple relationship between species richness and ecosystem processes is likely when one or a few species have strong ecosystem effects.

Although the relationship of species richness to ecosystem functioning has attracted considerable theoretical and experimental attention because of the irreversibility of species extinction, human activities influence the relative abundances of species more frequently than the presence or absence of species. Changes in species evenness warrant increased attention, because they usually respond more rapidly to human activities than do changes in species richness and because they have important consequences to ecosystems long before a species is threatened by extinction.

Species Composition

Particular species can have strong effects on ecosystem processes by directly mediating energy and material fluxes or by altering abiotic conditions that regulate the rates of these processes (Fig. 4)[13, 14]. Species' alteration of the availability of limiting resources, the disturbance regime, and the climate can have particularly strong effects on ecosystem processes. Such effects are most visible when introduced species alter previous patterns of ecosystem processes. For example, the introduction of the nitrogen-fixing tree *Myrica faya* to nitrogen-limited ecosystems in Hawaii led to a fivefold increase in nitrogen inputs to the ecosystem, which in turn changed most of the functional and structural properties of native forests[15]. Introduction of the deep-rooted salt cedar (*Tamarix* sp.) to the Mojave and Sonoran Deserts of North America increased the water and soil solutes accessed by vegetation, enhanced productivity, and increased surface litter and salts. This inhibited the regeneration of many native species, leading to a general reduction in biodiversity[16]. The perennial tussock grass, *Agropyron cristatum*, which was widely introduced to the northern Great Plains of North America after the 1930s 'dustbowl', has substantially lower allocation to roots compared with native prairie grasses. Soil under *A. cristatum*has lower levels of available nitrogen and ~25% less total carbon than native prairie soil, so the introduction of this species resulted in an equivalent reduction of 480 × 10^{12} g carbon stored in soils[17]. Soil invertebrates, such as earthworms and termites, also alter turnover of organic matter and nutrient supply, thereby influencing the species composition of the aboveground flora and fauna[18].

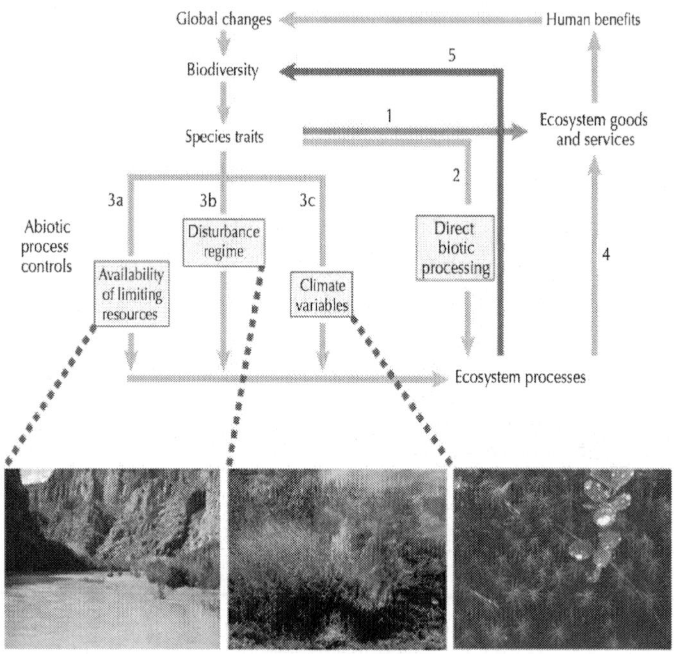

Figure 4: Mechanisms by which species traits affect ecosystem processes. Changes in biodiversity alter the functional traits of species in an ecosystem in ways that directly influence ecosystem goods and services (1) either positively (for example, increased agricultural or forestry production) or negatively (for example, loss of harvestable species or species with strong aesthetic/cultural value). Changes in species traits affect ecosystem processes directly through changes in biotic controls (2) and indirectly through changes

in abiotic controls, such as availability of limiting resources (3a), disturbance regime (3b), or micro- or macroclimate variables (3c). Illustrations of these effects include: reduction in river flow due to invasion of deep-rooted desert trees (3a; photo by E. Zavaleta); increased fire frequency resulting from grass invasion that destroys native trees and shrubs in Hawaii (3b, photo by C. D'Antonio); and insulation of soils by mosses in arctic tundra, contributing to conditions that allow for permafrost (3c; photo by D. Hooper). Altered processes can then influence the availability of ecosystem goods and services directly (4) or indirectly by further altering biodiversity (5), resulting in loss of useful species or increases in noxious species.

Species can also influence disturbance regime. For example, several species of nutritious but flammable grasses were introduced to the Hawaiian Islands to support cattle grazing. Some of these grasses spread into protected woodlands, where they caused a 300-fold increase in the extent of fire. Most of the woody plants, including some endangered species, are eliminated by fire, whereas grasses rebound quickly[19]. Similar increases in the ecological role of fire resulting from grass invasions have been widely observed in the Americas, Australia and elsewhere in Oceania. The invasion of cheatgrass (*Bromus tectorum*) into western North America is one of the most extensive of these invasions. Cheatgrass has increased fire frequency by a factor of more than ten in the >40 million hectares (1 ha = 10^4 m^2) that it now dominates[20].

Species-induced changes in microclimate can be just as important as the direct impacts of environmental change. For example, in late-successional boreal forests, where soil temperatures have a strong influence on nutrient supply and productivity, the presence of moss, which reduces heat flux into the soil, contributes to the stability of permafrost (frozen soils) and the characteristically low rates of nutrient cycling[21]. As fire frequency increases in response to high-latitude warming, moss biomass declines, permafrost becomes less stable, the nutrient supply increases, and the species composition of forests is altered. Plant traits can also influence climate at larger scales. Simulations with general circulation models indicate that widespread replacement of deep-rooted tropical trees by shallow-rooted pasture grasses would reduce evapotranspiration and lead to a warmer, drier climate[22]. At high latitudes, the replacement of snow-covered tundra by a dark conifer canopy will probably increase energy absorption sufficiently to act as a powerful positive feedback to regional warming[23].

Species Interactions

Most ecosystem processes are non-additive functions of the traits of two or more species, because interactions among species, rather than simple presence or absence of species, determine ecosystem characteristics (Fig. 5). Species interactions, including mutualism, trophic interactions (predation, parasitism and herbivory), and competition may affect ecosystem processes directly by modifying pathways of energy and material flow[24] or indirectly by modifying the abundances or traits of species with strong ecosystem effects[25].

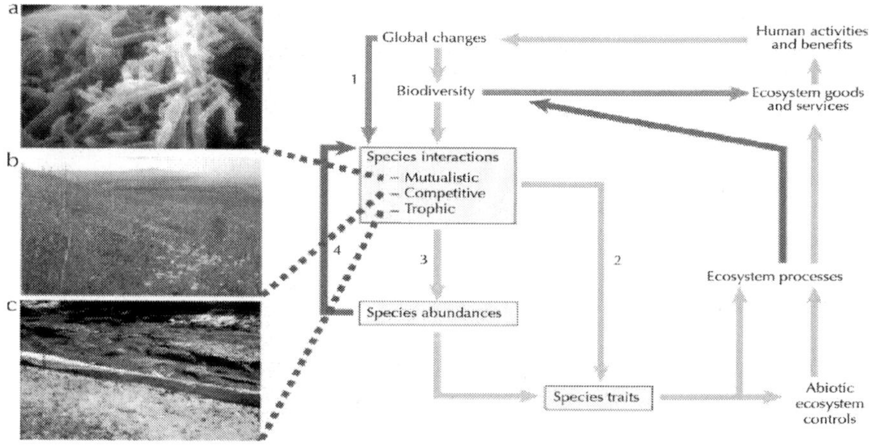

Figure 5: Mechanisms by which species interactions affect ecosystem processes. Global environmental change affects species interactions (mutualism, competition and trophic interactions) both directly (1) and through its effects on altered biodiversity. Species interactions may directly affect key traits (for example, the inhibition of microbial nitrogen fixation by plant secondary metabolites) in ecosystem processes (2) or may alter the abundances of species with key traits (3). Examples of these species interactions include (a) mutualistic consortia of microorganisms, each of which produces only some of the enzymes required to break down organic matter (photo by M. Klug), (b) altered abundances of native California forbs due to competition from introduced European grasses (photo by H. Reynolds), and (c) alteration of algal biomass due to presence or absence of grazing minnows[84] (photo by M. Power). Changes in species interactions and the resulting changes in community composition (3) may feedback to cause a cascade of further effects on species interactions (4).

Mutualistic species interactions contribute directly to many essential ecosystem processes. For example, nitrogen inputs to terrestrial ecosystems are mediated primarily by mutualistic associations between plants and nitrogen-fixing microorganisms. Mycorrhizal associations between plant roots and fungi greatly aid plant nutrient uptake from soil, increase primary production and speed succession[26]. Highly integrated communities (consortia) of soil microorganisms, in which each species contributes a distinct set of enzymes, speeds the decomposition of organic matter[27]. Many of these interactions have a high degree of specificity, which increases the probability that loss of a given species will have cascading effects on the rest of the system.

Trophic interactions can have large effects on ecosystem processes either by directly modifying fluxes of energy and materials, or by influencing the abundances of species that control those fluxes. When top predators are removed, prey populations sometimes explode and deplete their food resources, leading to a cascade of ecological effects. For example, removal of sea otters by Russian fur traders allowed a population explosion of sea urchins that overgrazed kelp[28] (Fig. 6a). Recent over-fishing in the North Pacific may have triggered similar outbreaks of sea urchin, as killer whales moved closer to shore and switched to sea otters as an alternate prey[29]. In the absence of dense populations of sea urchins, kelp provides the physical structure for diverse subtidal communities and attenuates waves that otherwise augment coastal erosion and storm damage[30]. Removing bass from lakes that were fertilized with phosphorus caused an increase in minnows, which depleted the biomass of phytoplankton grazers and caused algal blooms[31] (Fig. 6b). The algal blooms turned the lake from a net source to a net sink of CO_2. Thus, biotic change and altered nutrient cycles can interact to influence whole-system carbon balance. The zebra mussel (*Dreissena polymorpha*) is a bottom-dwelling invasive species that, through its filter feeding, markedly reduces phytoplankton while increasing water clarity and phosphorus availability[32]. Introduction of this species shifts the controlling interactions of the food web from the water column to the sediments. Trophic interactions are also important in terrestrial ecosystems. At the micro scale, predation on bacteria by protozoan grazers speeds nitrogen cycling near plant roots, enhancing nitrogen availability to plants[33]. At the regional scale, an improvement in hunting technology at the end of the Pleistocene may have contributed to the loss of the Pleistocene megafauna and the widespread change from steppe grassland to tundra that occurred in Siberia 10,000–18,000 years ago[34]. The resulting increase in mosses insulated the soil and led to cooler soils, less decomposition and greater sequestration of carbon in peat. Today, human harvest of animals continues to have a pronounced effect of the functioning of ecosystems.

Figure 6: Trophic interactions can affect ecosystem processes by influencing species' abundances. **a,** Removal of sea otters by Russian fur traders caused an explosion in the population of sea urchins that overgrazed kelp. (Photographs courtesy of M. Sewell/Still Pictures and J. Rotman/BBC Natural History Unit.) **b** , Similarly, changes in the species balance and the abundance of fish can deplete phytoplankton grazers and cause algal blooms. (Photograph courtesy of J. Foott/BBC Natural History Unit.)

Competition, mutualisms and trophic interactions frequently lead to secondary interactions among other species, often with strong ecosystem effects (Fig. 5). For example, soil microbial composition can modify the outcome of competition among plant species[35], and plants modify the microbial community of their neighbours, which, in turn, affects nitrogen supply and plant growth[36]. Stream predatory invertebrates alter the behaviour of their prey, making them more vulnerable to fish predation, which leads to an increase in the weight gain of fish[37]. In the terrestrial realm, grazers can reduce grass cover to the point that avian predators keep vole populations at low densities, allowing the persistence of *Erodium botrys*, a preferred food of voles[38]. The presence of *E. botrys*increases leaching[39] and increases soil moisture[40], which often limits production and nutrient cycling in dry grasslands. These examples clearly indicate that all types of organisms — plants, animals and microorganisms — must be considered in understanding the effects of biodiversity on ecosystem functioning. Although each of these examples is unique to a particular ecosystem, the ubiquitous nature of species interactions with strong ecosystem effects makes these interactions a general feature of ecosystem functioning. In many cases, changes in these interactions alter the traits that are expressed by species and therefore the effects of species on ecosystem processes. Consequently, simply knowing that a species is present or absent is insufficient to predict its impact on ecosystems.

Many global changes alter the nature or timing of species interactions[41]. For example, the timing of plant flowering and the emergence of pollinating insects differ in their responses to warming, with potentially large effects on ecosystems and communities[42]. Plant–herbivore interactions in diverse communities are less likely to be disrupted by elevated CO_2 (ref. 43) than in simple systems involving one specialist herbivore and its host plant[44].

Resistance and Resilience to Change

The diversity–stability hypothesis suggests that diversity provides a general insurance policy that minimizes the chance of large ecosystem changes in response to global environmental change[45]. Microbial microcosm experiments show less variability in ecosystem processes in communities with greater species richness[46], perhaps because every species has a slightly different response to its physical and biotic environment. The larger the number of functionally similar species in a community, the greater is the probability that at least some of these species will survive stochastic or directional changes in environment and maintain the current properties of the ecosystem[47]. This stability of processes has societal relevance. Many traditional farmers plant diverse crops, not to maximize productivity in a given year, but to decrease the chances of crop failure in a bad year[48]. Even the loss of rare species may jeopardize the resilience of ecosystems. For example, in rangeland ecosystems, rare species that are functionally similar to abundant ones become more common when grazing reduces their abundant counterparts. This compensation in response to release from competition minimizes the changes in ecosystem properties[49].

Species diversity also reduces the probability of outbreaks by 'pest' species by diluting the availability of their hosts. This decreases host-specific diseases[50], plant-feeding nematodes[51] and consumption of preferred plant species[52]. In soils, microbial diversity decreases fungal diseases owing to competition and interference among microbes[53].

Resistance to Invasions

Biodiversity can influence the ability of exotic species to invade communities through either the influence of traits of resident species or some cumulative effect of species richness. Early theoretical models and observations of invasions on islands indicated that species-poor communities would be more vulnerable to invasions because they offered more empty niches[54]. However, studies of intact ecosystems find both negative[55] and positive[56] correlations between species richness and invasions. This occurs in part because the underlying factors that generate differences in diversity (for example, propagule supply, disturbance regime and soil fertility) cannot be controlled and may themselves be responsible for differences in invasibility[56]. The diversity effects on invasibility are scale-dependent in some cases. For example, at the plot scale, where competitive interactions might exert their effect, increased plant diversity correlated with lower vulnerability to invasion in Central Plains grasslands of the United States. Across landscape scales, however, ecological factors that promote native plant diversity (for example, soil type and disturbance regime) also promote species invasions[57].

Experimental studies with plants[58] or soil microorganisms[59] often show that vulnerability to invasion is governed more strongly by the traits of resident and invading species than by species richness *per se*. Both competition and trophic interactions contribute to these effects of community composition on invasibility. For example, in its native range, the Argentine ant (*Linepithaema humile*) is attacked by species-specific parasitoids that modify its behaviour and reduce its ability to dominate food resources and competitively exclude other ant species[60]. These parasitoids are absent from the introduced range of Argentine ants, which may explain their success at eliminating native ant communities in North America[61]. Observational and experimental studies together indicate that the effect of species diversity on vulnerability to invasion depends on the components of diversity involved (richness, evenness, composition and species interactions) and their interactions with other ecological factors such as disturbance regime, resource supply and rate of propagule arrival. Humans significantly affect all of these factors (Figs 1, 4), thereby dramatically increasing the incidence of invasions worldwide.

SOCIETAL CONSEQUENCES OF ALTERED DIVERSITY

Biodiversity and its links to ecosystem properties have cultural, intellectual, aesthetic and spiritual values that are important to society. In addition, changes in biodiversity that alter ecosystem functioning have economic impacts through the provision of ecosystem goods and services to society (Fig. 1 and Box 2). Changes in diversity can directly reduce sources of food, fuel, structural materials, medicinals or genetic resources. These changes can also alter the abundance of other species that control ecosystem processes, leading to further changes in community composition and vulnerability to invasion. Introduction of exotic species or changes in community composition can affect ecosystem goods or services either by directly reducing abundances of useful species (by predation or competition), or by altering controls on critical ecosystem processes (Fig. 4).

These impacts can be wide-ranging and costly. For example, the introduction of deep-rooted species in arid regions reduces supplies and increases costs of water for human use. Marginal water losses to the invasive star thistle,*Centaurea solstitialis*, in the Sacramento River valley, California, have been valued at US$16–56 million per year (J. D. Gerlach, unpublished results) (Fig. 7). In South Africa's Cape region, the presence of rapidly transpiring exotic pines raises the unit cost of water procurement by nearly 30% (ref. 62). Increased evapotranspiration due to the invasion of *Tamarix* in the United States costs an estimated $65–180 million per year in reduced municipal and agricultural water supplies[63]. In addition to raising water costs, the presence of sediment-trapping*Tamarix* stands has narrowed river channels and obstructed over-bank flows throughout the western United States, increasing flood damages by as much as $50 million annually[63].

Figure 7: Water losses to the invasive, deep-rooted star thistle, C. solstitialis , provides an example of the financial impacts of introducing exotic species on ecosystem composition.

(Photograph courtesy of P. Collins/A-Z Botanical Collection.).

Those species changes that have greatest ecological impact frequently incur high societal costs. Changes in traits maintaining regional climate[22] constitute an ecosystem service whose value in tropical forests has been estimated at $220 ha^{-1} yr^{1} (ref. 64). The loss or addition of species that alter disturbance regimes can also be costly.

The increased fire frequency resulting from the cheatgrass invasion in the western United States has reduced rangeland values and air quality and led to increased expenditures on fire suppression[65]. The disruption of key species interactions can also have large societal and ecological consequences. Large populations of passenger pigeons (*Ectopistes migratorius*) in the northeastern United States may once have controlled Lyme tick-bearing mice by out-competing them for food[66]. The loss of the passenger pigeon to nineteenth-century over-hunting may, therefore, have contributed to the rise of Lyme disease in humans in the twentieth century. The economic impacts of invasions of novel species are particularly well documented. The introduction and spread of single pests such as the golden apple snail (*Pomacea canaliculata*) and the European corn borer (*Ostrinia nubilalis*) have had major impacts on food production and farm incomes[67, 68]. Estimates of the overall cost of invasions by exotic species in the United States range widely from $1.1 to $137 billion annually[69, 70]. In Australia, plant invasions alone entail an annual cost of US$2.1 billion[71].

The provision of tangible ecosystem goods and services by natural systems depends not only on species' presence or absence but also on their abundance. Large populations of the white-footed mouse (*Peromyscus leucopus*) in the northeastern United States control outbreaks of gypsy moth (*Lymatria dispar*) but spread Lyme disease, whereas small populations of the mouse decrease the incidence of Lyme disease but allow gypsy moth defoliation[72]. An analysis of the costs of changes in biodiversity thus involves more than just analysis of extinctions and invasions. The loss of a species to extinction is of special societal concern, however, because it is irreversible. Future opportunities to learn and derive newly recognized benefits from an extinct species are lost forever. Preventing such a loss preserves an 'option value' for society — the value of attaining more knowledge about species and their contribution to human well being in order to make informed decisions in the future[73, 74]. For example, significant value ($230–330 million) has been attributed to genetic information gained from preventing land conversion in Jalisco, Mexico, in an area containing a wild grass, teosinte (*Euchlaena mexicana*), that can be used to develop viral-resistant strains of perennial corn[73]. If this land had been converted to agriculture or human settlements, the societal benefits of development would have come at the expense of an irreversible loss in genetic material that could be used for breeding viral resistance in one of the most widely consumed cereal crops in the world. The perceived costs of diversity loss in this situation might have been small — especially relative to the development benefits — whereas the actual (unrecognized) costs of losing genetic diversity would have been significant (Fig. 8). Decisions to preserve land to gain further information about the societal value of species diversity or ecosystem function typically involve a large degree of uncertainty, which often leads to myopic decisions regarding land use.

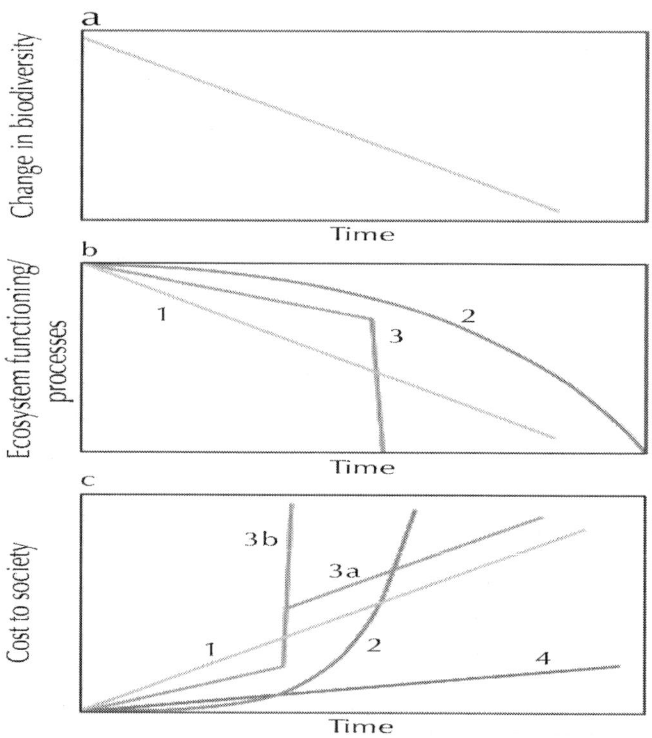

Figure 8: Ecosystem and societal consequences of changes in biodiversity. **a**, A linear change in biodiversity through time. **b**, This change might (1) induce a linear response in ecosystem processes, (2) have increasingly large impacts on ecosystem functioning, yielding exponential ecosystem change through time, or (3) exhibit abrupt thresholds owing to the loss of a keystone species, the loss of the last member of a key functional group, or the addition of a new species trait. **c**, Even if ecosystem response to diversity changes is linear, associated societal costs through time may respond nonlinearly. Departures

from a linear increase (1) in societal costs over time might include larger cost increases (2) associated with each additional unit of change in ecosystem processes, yielding an exponential cost curve through time. Reductions of resource supply below threshold levels may induce step increases in societal costs (3a), such as reductions in water supply below the point where all consumers have access to enough for desired uses. If changes in resource supply or ecosystem processes exceed thresholds for supporting large segments of society, stepwise cost increases may be unmeasurable or essentially infinite (3b). The perceived ecological changes and societal costs of diversity change may be small (4). Actual, unrecognized costs may be far higher (lines 1, 2 and 3) and discovered only later as lost option values. Conservation of biodiversity can help avoid such negative ecological and economic 'surprises'.

Global environmental changes have the potential to exacerbate the ecological and societal impacts of changes in biodiversity[6]. In many regions, land conversion forces declining populations towards the edges of their species range, where they become increasingly vulnerable to collapse if exposed to further human impact[75]. Warming allows the poleward spread of exotics and pathogens, such as dengue- and malaria-transmitting mosquitoes (*Aedes* and*Anopheles* sp.)[76] and pests of key food crops, such as corn-boring insects[68]. Warming can also exacerbate the impacts of water-consuming invasive plant species in water-scarce areas by increasing regional water losses. The *Tamarix*-invaded Colorado River in the United States currently has a mean annual flow that is 10% less than regional water allocations for human use[77]. Warming by 4°C would reduce the flow of the Colorado River by more than 20%, further increasing the marginal costs of water losses to *Tamarix*[78]. Similar impacts of global change in regions such as Sahelian Africa, which have less water and less well developed distribution mechanisms, might directly affect human survival. In many cases, accelerated biodiversity loss is already jeopardizing the livelihoods of traditional peoples[79].

The combination of irreversible species losses and positive feedbacks between biodiversity changes and ecosystem processes are likely to cause nonlinear cost increases to society in the future, particularly when thresholds of ecosystem resilience are exceeded[80]. For example, *Imperata cylindrica* , an aggressive indigenous grass, colonizes forest lands of Asia that are cleared for slash-and-burn agriculture, forming a monoculture grassland with no vascular plant diversity and many fewer mammalian species than the native forest.

The total area of *Imperata* in Asia is currently about 35 million ha (4% of land area)[81]. Once in place, *Imperata* is difficult and costly to remove and enhances fire, which promotes the spread of the grass. The annual cost of reversing this conversion in Indonesia, where 4% of the nation's area (8.6 million ha) is now in*Imperata* grasslands, would be over $400 million if herbicides are used, and $1.2 billion if labour is used to remove the grass manually. Farmers typically burn the fields because herbicides and labour are too expensive. Burning these grasslands, however, increases losses of soil nitrogen and carbon, which erode agricultural productivity, and enhances regeneration of *Imperata*. This positive feedback with nonlinear changes in land cover will probably continue in the future as lands are deforested for timber and agricultural purposes, causing further declines in regional biodiversity.

Uncertainty related to positive feedbacks and nonlinear changes in land cover and biodiversity make social adaptation to change more difficult and costly (Fig. 8). It may be more important from an economic perspective to understand the nature and timing of rapid or nonlinear changes in societal costs caused by loss of biodiversity and associated ecosystem services than it is to predict average consequences of current trends of species decline. By analogy, economic models of ecological 'surprises' in response to climatic change show that the information about the nonlinearities in damage from warming is worth up to six times more than information about current trends in damage levels[82]. In the *Imperata*example, the costs of replacing the original ecosystem goods and services from the forest — including timber products, fire stability and soil nutrients — rise sharply as *Imperata* spreads. If these nonlinearities in the ecological and economic effects of this conversion had been anticipated, policies could have been implemented to encourage agroforestry instead of rice production or to reduce migration and settlement in the most vulnerable areas[83].

In sum, these examples indicate a tight coupling between altered species diversity, ecosystem function and societal costs. A pressing task for ecologists, land managers and environmental policy makers is to determine where and when such tight couplings exist. Policies to safeguard ecosystem services must be able to respond dynamically to new knowledge, the rapidly changing global environment, and evolving societal needs. Nonlinearity, uncertainty and irreversibility call for a more aggressive approach to mitigating changes in biodiversity than is now being pursued so that future options are not foreclosed.

CONCLUSIONS

We are in the midst of one of the largest experiments in the history of the Earth. Human effects on climate, biogeochemical cycles, land use and mobility of organisms have changed the local and global diversity of the planet, with important ecosystem and societal consequences (Fig. 1). The most important causes of altered biodiversity are factors that can be regulated by changes in policy: emissions of greenhouse gases, land-use change and species introductions. In the past, the international community has moved to reduce detrimental human impacts with unambiguous societal consequences. For example, the Montreal Protocol prohibited release of chlorofluorocarbons in response to evidence that these chemicals caused loss of ozone and increased levels of cancer-producing UV-B radiation. Strong evidence for changes in biodiversity and its ecosystem and societal consequences calls for similar international actions. We urge the following blueprint for action.

- The scientific community should intensify its efforts to identify the causes of nonlinearities and thresholds in the response of ecosystem and social processes to changes in biodiversity.
- The scientific community and informed citizens should become engaged in conveying to the public, policy-makers and land managers the enormity and irreversibility of current rapid changes in biodiversity. Despite convincing scientific evidence, there is a general lack of public awareness that change in biodiversity is a global change with important ecological and societal impacts and that these changes are not amenable to mitigation after they have occurred.
- Managers should consider the ecological and social consequences of biodiversity change at all stages in land-use planning. For example, environmental impact assessments should consider both the current costs of ecosystem services that will be lost and the risk of nonlinear future change. Managed landscapes can support a large proportion of regional biodiversity with proper planning, management and adaptive responses.
- Scientists and other citizens should collaborate with governmental organizations, from local to national levels, in developing and implementing policies and regulations that reduce environmental

deterioration and changes in biodiversity. For example, more stringent restrictions on the import of biotic materials could curb the rate of biotic invasions, and improved land and watershed management could reduce their rates of spread.

- A new international body that would be comparable to the Intergovernmental Panel on Climate Change (IPCC) should assess changes in biodiversity and their consequences as an integral component of the assessment of the societal impacts of global change.
- International bodies should establish and implement agreements such as the Convention on Biological Diversity that institute mechanisms for reducing activities that drive the changes in biodiversity. These activities include fossil-fuel emissions, land-use change and biotic introductions.

ACKNOWLEDGEMENTS

We thank B. R. Tershy for valuable inputs and J. D. Gerlach for access to his unpublished manuscript.

REFERENCES

1. Postel, S. L., Daily, G. C. & Ehrlich, P. R. Human appropriation of renewable fresh water. *Science* 271, 785–788 (1996).
2. Vitousek, P. M., Mooney, H. A., Lubchenco, J. & Melillo, J. M. Human domination of Earth's ecosystems. *Science* 277 , 494–499 (1997).
3. Kattenberg, A. *et al.* in *Climate Change 1995. The Science of Climate Change*(ed. Houghton, J. T.) 285–357 (Cambridge Univ. Press, Cambridge,1996).
4. Pimm, S. L., Russell, G. J., Gittleman, J. L. & Brooks, T. M. The future of biodiversity. *Science* 269, 347–350 (1995).
5. Lawton, J. H. & May, R. M. *Extinction Rates* (Oxford Univ.Press, Oxford,1995).
6. Sala, O. E. *et al.* Global biodiversity scenarios for the year 2100. *Science*287, 1770–1776 (2000).

7. Lawton, J. H. What do species do in ecosystems? *Oikos* 71, 367–374 (1994). | ISI |

8. Tilman, D., Wedin, D. & Knops, J. Productivity and sustainability influenced by biodiversity in grassland ecosystems. *Nature* 379, 718 –720 (1996).

9. Hector, A. *et al.* Plant diversity and productivity experiments in European grasslands . *Science* 286, 1123–1127 (1999).

10. van der Heijden, M. G. A. *et al.* Mycorrhizal fungal diversity determines plant biodiversity, ecosystem variability and productivity. *Nature* 396, 69–72 (1998).

11. Salonius, P. O. Metabolic capabilities of forest soil microbial populations with reduced species diversity. *Soil Biol. Biochem.* 13, 1– 10 (1981).

12. Wardle, D. A., Bonner, K. I. & Nicholson, K. S. Biodiversity and plant litter: experimental evidence which does not support the view that enhanced species richness improves ecosystem function. *Oikos* 79, 247– 258 (1997).

13. Hooper, D. U. & Vitousek, P. M. The effects of plant composition and diversity on ecosystem processes. *Science* 277, 1302–1305 (1997).

14. Tilman, D. *et al.* The influence of functional diversity and composition on ecosystem processes. *Science* 277, 1300– 1302 (1997).

15. Vitousek, P. M., Walker, L. R., Whiteaker, L. D., Mueller-Dombois, D.& Matson, P. A. Biological invasion by *Myrica faya* alters ecosystem development in Hawaii. *Science* 238, 802–804 (1987).

16. Berry, W. L. Characteristics of salts secreted by *Tamarix aphylla*. *Am. J. Bot.* 57, 1226–1230 (1970).

17. Christian, J. M.& Wilson, S. D. Long-term impacts of an introduced grass in the northern Great Plains. *Ecology* 80, 2397–2407 (1999).

18. Lavelle, P., Bignell, D. & Lepage, M. Soil function in a changing world: the role of invertebrate ecosystem engineers. *Eur. J. Soil Biol.* 33, 159–193 (1997).

19. D'Antonio, C. M.& Vitousek, P. M. Biological invasions by exotic grasses, the grass-fire cycle, and global change. *Annu. Rev. Ecol. Syst.* 23, 63–87(1992).

20. Whisenant, S. *Changing Fire Frequencies on Idaho's Snake River Plains: Ecological Management Implications* 4–10 (US Forest Service GeneralTechnical Report INT-276, Washington, 1990).
21. Van Cleve, K., Chapin, F. S. III, Dryness , C. T. & Viereck, L. A. Element cycling in taiga forest: state-factor control . *BioScience* 41, 78–88(1991). | Article |
22. Shukla, J., Nobre, C. & Sellers, P. Amazon deforestation and climate change. *Science* 247, 1322–1325 (1990).
23. Foley, J. A., Kutzbach, J. E., Coe, M. T. & Levis, S. Feedbacks between climate and boreal forests during the Holocene epoch. *Nature* 371, 52–54 (1994).
24. de Ruiter, P. C., Neutel, A. & Moore, J. C. Energetics, patterns of interaction strengths, and stability in real ecosystems. *Science* 269, 1257– 1260 (1995).
25. Power, M. E. *et al*. Challenges in the quest for keystones. *BioScience* 46, 609 (1996).
26. Read, D. J. Mycorrhizas in ecosystems. *Experientia* 47, 376–391 (1991).
27. Paerl, H. W. & Pinckney, J. L. A mini-review of microbial consortia: their roles in aquatic production and biogeochemical cycling. *Microbial Ecol.* 31, 225–247 (1996).
28. Estes, J. A. & Palmisano, J. F. Sea otters: their role in structuring nearshore communities. *Science* 185, 1058 –1060 (1974).
29. Estes, J. A., Tinker, M. T., Williams, T. M. & Doak, D. F. Killer whale predation on sea otters linking oceanic and nearshore ecosystems .*Science* 282, 473–476(1998).
30. Mork, M. The effect of kelp in wave damping. *Sarsia* 80, 323–327 (1996). | ISI |
31. Schindler, D. E., Carpenter, S. R., Cole, J. J., Kitchell, J. F. & Pace, M. L.Influence of food web structure on carbon exchange between lakes and the atmosphere. *Science* 277, 248–251 (1997).
32. Caraco, N. F. *et al*. Zebra mussel invasion in a large, turbid river: phytoplankton response to increased grazing. *Ecology* 78, 588–602 (1997).
33. Clarholm, M. Interactions of bacteria, protozoa and plants leading to mineralization of soil nitrogen. *Soil Biol. Biochem.* 17, 181 –187 (1985).

34. Zimov, S. A. *et al.* Steppe-tundra transition: an herbivore-driven biome shift at the end of the Pleistocene *Am. Nat.* 146, 765–794 (1995).

35. Chanway, C. P., Turkington, R. & Holl, F. B. Ecological implications of specificity between plants and rhizosphere micro-organisms. *Adv. Ecol. Res.* 21 , 121–169 (1991).

36. Lawley, R. A., Newman, E. I. & Campbell, R. Abundance of endomycorrhizas and root-surface microorganisms on three grasses grown separately and in mixtures. *Soil Biol. Biochem.* 14, 237–240 (1982).

37. Soluck, D. A. & Richardson, J. S. The role of stoneflies in enhancing growth of trout: a test of the importance of predator–predator facilitation within a stream community. *Oikos* 80, 214–219 (1997).

38. Rice, K. J. Interaction of disturbance patch size and herbivory in *Erodium*colonization . *Ecology* 68, 1113–1115 (1987).

39. Shock, C. C., Jones, M. B., Williams, W. A. & Center, D. M. Competition of S and N by associations of three annual range species in lysimeters *Plant Soil* 81, 311–321 (1984).

40. Gordon, D. R. & Rice, K. J. Partitioning of space and water between two California annual grassland species. *Am. J. Bot.* 79, 967–976 (1992).

41. Jifon, J. L., Friend, A. L. & Berrang, P. C. Species mixture and soil-resource availability affectthe root growth response of tree seedlings to elevated atmospheric CO_2. *Can. J. For. Res.* 25, 824–832 (1995).

42. Harrington, R., Woiwod, I. & Sparks, T. Climate change and trophic interactions. *Trends Ecol. Evol.* 14, 146–150(1999).

43. Díaz, S., Fraser, L. H., Grime, J. P. & Falczuk, V. Theimpactofelevated CO_2 on plant–herbivore interactions: experimentalevidence of moderating effects at the community level. *Oecologia* 117, 177–186 (1998).

44. Lindroth, R. L.in *Carbon Dioxide and Terrestrial Ecosystems* (eds Koch, G. W. &Mooney, H. A.) 105–120 (Academic Press, San Diego, 1996).

45. McNaughton, S. J. Diversity and stability of ecological communities: a comment on the role of empiricism in ecology. *Am. Nat.* 111, 515 –525 (1977).

46. Naeem, S. & Li, S. Biodiversity enhances ecosystem reliability . *Nature*390, 507–509(1997).
47. Chapin, F. S. III & Shaver, G. R. Individualistic growth response of tundra plant species to environmental manipulations in the field. *Ecology* 66, 564– 576 (1985).
48. Altieri, M. A.in *Agroecology* (eds Carrol, C. R., Vandermeer, J. H. & Rosset,P. M.) 551–564 (McGraw Hill, New York, 1990).
49. Walker, B., Kinzig, A. & Langridge, J. Plant attribute diversity, resilience, and ecosystem function: the nature and significance of dominant and minor species. *Ecosystems* 2, 95–113 (1999).
50. Burdon, J. J. The structure of pathogen populations in natural plant communities. *Annu. Rev. Phytopathol.* 31, 305–323 (1993).
51. Wasilewska, L. Differences in development of soil nematode communities in single- and multi-species grass experimental treatments. *Appl. Soil Ecol.* 2, 53–64 (1995).
52. Bertness, M. D.& Leonard, G. H. The role of positive interactions in communities: lessons from intertidal habitats. *Ecology* 78, 1976–1989 (1997).
53. Nitta, T. Diversity of root fungal floras: its implications for soil-borne diseases and crop growth. *Jap. Agric. Res. Quart.* 25, 6–11 (1991).
54. Elton, C. S. *The Ecology of Invasions by Animals and Plants* (Methuen, London, 1958).
55. Tilman, D. Community invasibility, recruitment limitation, and grassland biodiversity . *Ecology* 78, 81–92(1997).
56. Levine, J. M. & D'Antonio, C. M. Elton revisited: a review of evidence linking diversity and invasibility. *Oikos* 87, 15–26 (1999).
57. Stohlgren, T. J. *et al.* Exotic plant species invade hot spots of native plant diversity . *Ecol. Monogr.* 69, 25– 46 (1999).
58. Lavorel, S., Prieur-Richard, A.-H & Grigulis, K. Invasibility and diversity of plant communities: from patterns to processes. *Diversity Distrib.* 5, 41–49 (1999).
59. McGrady-Steed, J., Harris, P. M. & Morin, P. J. Biodiversity regulates ecosystem predictability. *Nature* 390, 162–165 (1997).

60. Orr, M. R. & Seike, S. H. Parasitoids deter foraging by Argentine ants (*Linepithema humile*) in their native habitat in Brazil. *Oecologia* 117, 420–425 (1998).

61. Holway, D. A. Competitive mechanisms underlying the displacement of native ants by the invasive Argentine ant. *Ecology* 80, 238– 251 (1999).

62. Van Wilgen, B. W., Cowling, R. M. & Burgers, C. J. Valuation of ecosystem services: a case study from South African fynbos ecosystems. *BioScience*46, 184–189 (1996).

63. Zavaleta, E. S.in *Invasive Species in a Changing World* (eds Hobbs, R. J. &Mooney, H. A.) (Island, Washington DC, in the press).

64. Costanza, R. *et al*. The value of the world's ecosystem services and natural capital . *Nature* 387, 253–260(1997).

65. Stewart, G. & Hull, A. C. Cheatgrass (*Bromus tectorum* L.) — an ecological intruder in southern Idaho. *Ecology* 30, 58–74 (1949).

66. Blockstein, D. E. Letter to the editor. *Science* 279, 1831(1998).

67. Naylor, R. L. Invasions in agriculture: assessing the cost of the golden apple snail in Asia. *Ambio* 25, 443–448 (1996).

68. Porter, J. H., Parry, M. L. & Carter, T. R. The potential effects of climatic change on agricultural insect pests. *Agric. For. Meteorol.* 57, 221–240 (1991).

69. Pimentel, D., Lach, L., Zuniga, R.& Morrison, D. Environmental and economic costs of nonindiginous species in the United States. *BioScience*50, 53–65 (2000).

70. USOT Assessment. *Harmful Non-indigenous Species in the United States*(US Government Printing Office, Washington, 1993).

71. Thorp, J. *The National Weeds Strategy: A Strategic Approach to Weed Problems of National Significance* (Agriculture and Resource Management Council of Australiaand New Zealand, Canberra, 1997).

72. Jones, C. G., Ostfeld, R. S., Richard, M. P., Schauber, E. M.& Wolff, J. O.Chain reactions linking acorns to gypsy moth outbreaks and Lyme disease risk. *Science* 279, 1023–1026 (1998).

73. Fisher, A. C. & Hanemann, W. M. Option value and the extinction of species. *Adv. Appl. Micro-Econ.* 4, 169 –190 (1986).

74. Naylor, R. L. in *Invasive Species in a Changing World* (eds Hobbs, R. & Mooney, H.A.) (Island, Washington DC, in the press).

75. Channell, R. & Lomolino, M. V. Dynamic biogeography and conservation of endangered species. *Nature* 403, 84– 86 (2000).

76. Bryan, J. H., Foley, D. H. & Sutherst, R. W. Malaria transmission and climate change in Australia . *Med. J. Aust.* 164, 345– 347 (1996).

77. Morrison, J. I. *The Sustainable Use of Water in the Lower Colorado River Basin* (PacificInstitute and the Global Water Policy Project, Oakland, 1996).

78. Zavaleta, E. S. The emergence of waterfowl conservation among Yup'ik hunters in the Yukon-Kuskokwim Delta, Alaska. *Hum. Ecol.* 27, 231– 266 (2000).

79. Warren, D. M. & Pinkston, J. in *Linking Social and Ecological Systems* (eds Berkes, F. & Folke, C.) 158– 189 (Cambridge Univ. Press, Cambridge,1998).

80. Schlesinger, W. H. *et al.* Biological feedbacks in global desertification.*Science* 247, 1043–1048 (1990).

81. Garrity, D. P. *et al.* The *Imperata* grasslands of tropical Asia: area, distribution, and typology. *Agrofor. Syst.* 36, 1– 29 (1997).

82. Peck, S. C. &Teisberg, T. J. in *Assessing Surprises and Nonlinearities in Greenhouse Warming* (eds Darmstadter, J. & Toman, M. A.) 80–83 (Resources for the Future, Washington, 1993).

83. Tomich, T. P., Kuusipalo, J., Menz, K.& Byron, N. *Imperata* economics and policy. *Agrofor. Syst.* 36, 233–261 (1997).

84. Power, M. E., Matthews, W. J. & Steward, A. J. Grazing minnows, piscivorous bass, and stream algae: dynamics of a strong interaction.*Ecology* 66, 1448–1456 (1985).

85. Johnson, K. G., Vogt, K. A., Clark, H. J. , Schmitz, O. J. & Vogt, D. J.Biodiversity and the productivity and stability of ecosystems. *Trends Ecol. Evol.* 11, 372–377 (1996).

86. Vitousek, P. M.& Hooper, D. U. in *Biodiversity and Ecosystem Function*(eds Schulze, E.-D & Mooney, H. A.) 3–14 (Springer, Berlin, 1993).

87. Huston, M. A. Hidden treatments in ecological experiments: re-evaluating the ecosystem function of biodiversity. *Oecologia* 110, 449– 460 (1997). | Article | ISI |

88. Daily, G. C. *Nature's Services: Societal Dependence on Natural Ecosystems* (Island,Washington DC, 1997).

89. Goulder, L. H.& Kennedy, D. in *Nature's Services: Societal Dependence on Natural Ecosystems* (ed. Daily, G. C.) 23 –48 (Island, Washington DC,1997).

Chapter 4

Seeking Common Ground: How Natural and Social Scientists Might Jointly Create an Overlapping Worldview for Sustainable Livelihoods: A South African Perspective

Nick King[1], Harry Biggs[2], Rael Loon[3]

[1]Endangered Wildlife Trust, South Africa

[2]Scientific Services, South African National Parks, South Africa

[3]Transboundary Protected Area Research Initiative/IUCN-SA, South Africa

ABSTRACT

In this paper, an attempt is made to identify key factors that may enhance the basis for collaboration at the interface between social and natural sciences, and to describe conceptually how natural capital differs from social capital. The paper begins by building on what is believed to be concepts common to both fields. In particular, it equates ecosystem services with natural capital, drawing especially on notions of the underpinning nature of ecosystem services as natural capital and suggesting key aspects of social science that add indispensable value to conservation, as part of a wider recognition of social capital. It is concluded that it is most realistic and productive to regard the world, not as compartmentalised, but as an integrated socioecological system. In this regard by way of example, the positive potential of ecotourism in South Africa and how it integrates natural and social capital is contrasted with the negative potential of Invasive Alien Species. This leads to a discussion of South Africa's National Biodiversity Framework as an opportunity to make use of an integrated approach to biodiversity conservation. The mutually beneficial co-existence of people and protected areas is pivotal to any success in this regard and the 'Swi ta Lunga' Trust is discussed as an emerging Integrated Conservation and Development project which has potential for mutually reinforcing positive natural and social capitals. A common understanding of the interconnectedness and interdependence of biophysical and social sciences, is the key to creating a joint platform for biophysical and social scientists, helping them to develop common objectives and a unified, or at least a significantly overlapping worldview. The paper concludes that a collaborative effort is required to drive the 'strong sustainability' paradigm that will allow the attainment of this common objective.

INTRODUCTION

The Term Globalisation in the twenty-first century has many meanings, but, without doubt, we are seeing much closer interactions between various human endeavours, from commerce to communications, scientific disciplines to social interactions, all facilitated by improved information and communication technologies. The biophysical and social sciences are no exception. Contemporary society can be likened to a 'global village' with a synthesis of diverse multidisciplinary and interdisciplinary issues taking place. While the study of isolated systems will always be valuable, there is an increasing trend to address cross-disciplinary issues (Lele and Norgaard 2005).

However, as with many such interactions, the first forays are often fraught with misunderstandings and attempts to impose respective ideals on the other party, each party often having differing underlying paradigms and resultant contrasting worldviews. This tends to lead to a simultaneous withdrawal from the 'frontline' and a rehashing of long-held perspectives that the opposing side is incapable of broadening its mindset, and thus will be impossible to work with. Recognising this and working together in an interdisciplinary manner to find common ground and a common language will help to bridge these divides. This, in turn, will provide the basis for an improved and increasingly integrated understanding of both worldviews. Conservation that is intricately linked with society and socio-economic concerns (such as addressing the needs of local communities adjacent to protected areas) is increasingly becoming pivotal to biodiversity conservation planning. Especially in megabiodiverse countries such as South Africa, such an approach may facilitate new and improved solutions to this current social-conservation conundrum and the need for effectively integrating conservation and social concerns (see, e.g. Curtin 2002; Magome and Fabricius 2004; Lele and Norgaard 2005).

From a natural science perspective such a conundrum is evident in the limited quantification of social aspects of conservation (such as measuring community participation for instance), which often makes it difficult to reconcile social sciences with the natural science paradigm. Natural and social scientists also seem to have differing basic assumptions about each other, which make it difficult for both to be incorporated on the same basis.

For example, a common (and often credible) caricature of a natural (biophysical) scientist is one who only accepts something that has been through the hypothetico-deductive filter of Popperian falsification and thus qualifies as 'hard science'. One side of our argument in this paper is that natural scientists may be overconfident as to how successful their 'hard' knowledge will be in the real world, and that this 'hard knowledge' often fails because they have not considered the appropriate complementary social 'soft' science. This is often more difficult to come by and once come by it is difficult to use, hence the dictum 'the hard part is the soft part'. While ecologists are not considered hard scientists in the strictest sense, the perception is still that they are 'harder' scientists than their social counterparts.

Conversely, natural scientists sometimes contend that social scientists view the world as one in which there are no immutable laws that cannot be circumvented; more importantly perhaps that the human condition is not subject to the laws of nature, at least in the short term, but to the vagaries of human political power disparities. In this worldview, human influences override natural influences, and it is human relationships that determine the human condition. Some natural scientists contend that such a standpoint lacks a realistic appreciation of the underpinning natural sciences by simply discounting environmental concerns and management as socially dependent, often failing to recognise the limited supply of natural resources, and the concept of limits to the biophysical world in general.

Thus, the start of our common understanding, we opine, is to see the environment as not only physically, but also socially and politically constructed (Brandon and Wells 1992; Brandon et al. 1998). The key here is that this is not an either/or scenario, as exponents in either camp often claim-the interconnectedness of the two, biophysical and social, rather than either component, is the missing link. Of the three, it is this last, the interdependency that is the critical component.

The first aim of this paper is to identify which integrating disciplines are relevant for the purposes of this discussion. Then we present a simple model using the notions of natural and social capital to illustrate the relationship between the natural and social sciences and to describe how good vs poor governance structures can lead to either sustainable or unsustainable solutions. This in turn may have either positive or negative implications for human livelihoods. These concepts are applied to two case studies in order to illustrate a sustainable solution where capitals may be growing (the ecotourism industry in South Africa) vs an unsustainable solution where capitals may have declined [Invasive Alien Species (IAS)]. Recommendations are made as to how an integrated approach adopted in the context of the South African National Biodiversity Framework might enhance a strong sustainable scenario. Finally, we discuss how such 'good' (i.e.

positively reinforcing) natural and social capitals may be effectively harmonised and mutually reinforced in Integrated Conservation and Development Projects that address a number of 'people and parks' issues.

Therefore, the aims of this paper are: (1) to explain the similarities and differences in the perspectives of the natural and social sciences and suggest how some of these different fields can be effectively integrated; (2) to present a framework to explain the differences between natural and social capital in the context of good or bad governance and how this impinges on sustainability and peoples' livelihoods and (3) to explore the dynamics of such concepts using two relevant South African case studies and suggest how an integrated approach may facilitate promising directions such as the formulation of the South African National Biodiversity Framework as initiated by recent legislation.

BRIDGING THE DIVIDE

Although it seems that we are on a road towards globalisation, the modern world is still characterised by an unprecedented fragmentation and specialisation of knowledge, including scientific knowledge. Yet solving environmental problems requires that natural and social scientists bring together such dispersed knowledge to inform collective deliberation over the issue in question (Mascia et al. 2003). To build such common ground we first highlight those broad areas and disciplines unique to the natural sciences and the social sciences (relevant to this paper) and propose some areas we think might help to provide a joint understanding of the situation. We then suggest an 'intermediary' framework for working together, based on cardinal aspects that impact on the natural environment.

Traditionally, mainstream biophysical scientists tend to see the world as set within a framework of rigorous natural laws [see for example Heinberg (2003)], for an in-depth analysis of the finite nature of energy available to us both globally and locally.

Natural sciences tend to have more defined parameters and are rigorous in the sense of being more limited and constrained than the social sciences. Natural science fields that broadly contribute to natural resource management include ecology, environmental sciences, conservation biology and agricultural science. Social science fields that broadly impinge on the environment include political science, law, economics, rural development and sociology (see [Table 1]).

Table 1: Integrating various social and natural science fields as contemporary interdisciplinary tools for environmental management relevant to this paper

	Social science discipline	Integrating discipline	Natural science discipline
1	Political science	Environmental politics and policy analysis	Environmental sciences
2	Public policy and management	Integrated environmental management	Environmental sciences
3	Law	Environmental law	Environmental sciences
4	Rural development	Community-based natural resource management	Natural resource management
5	Rural development	Integrated conservation and development	Conservation biology
6	Tourism and hospitality	Sustainable tourism	Conservation./ natural resource management
7	Economics	Ecological economics	Ecology
8	Economics	Resource economics	Natural resource management
9	Economics	Conservation economics	Conservation biology
10	Economics	Agricultural economics	Agricultural sciences
11	Sociology/ humanities	Social anthropology; archaeology	Physical anthropology
12	Sociology/ humanities	Human/social/ political ecology	Ecology

Certain narrower overlapping disciplines that combine both natural and social sciences and which may match issues relevant to each are listed in [Table 1]. Achieving sustainable use of resources we argue, very broadly requires cross-disciplinary work at the confluence of social, economic and ecological fields (see [Figure 1]). We provide two case studies selected to reflect real-world examples of either side of the social-ecology feedback framework and also how natural and social science can be integrated. We argue that such a combined approach can make positive contributions to a number of societal and conservation problems and issues: the ecotourism case study demonstrates how an understanding of sustainable tourism is needed to manage this substantial industry in South Africa both at the policy and local levels; the case study on IAS provides an example in ecological economics by demonstrating the negative ecological and economic cost of poor governance and the need for stronger environmental management policies and practices.

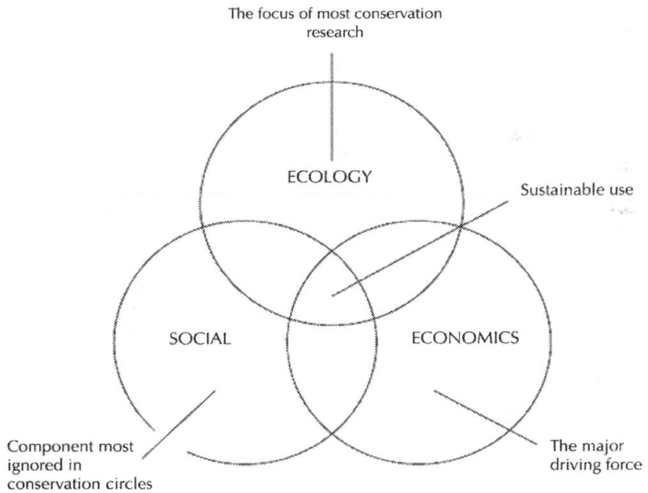

Figure 1: Achieving sustainable use of resources requires cross-disciplinary work at the confluence of social, economic and ecological fields.

Another example of an integrated approach can be found in the field of environmental law, which covers prescriptive issues with both natural and social components. South Africa has recently seen the emergence of progressive environmental legislation (such as the Protected Areas

Act and Biodiversity Act), and environmental law provides one sound basis to address such issues. The field of 'ecological economics' may also provide useful tools for tackling such issues with both ecological and economic components. Since many contemporary real-world problems have both social and ecological dimensions constituted by natural capital and human/social capital, this field views the economic system as being embedded in the ecological-social system (see [Figure 2]). As such, ecological economics may provide another sound theoretical basis to address such issues (Aronson et al 2007). These two fields, i.e. environmental law and ecological economics are used to propose a 'stick and carrot' strategy relating to the National Biodiversity Framework later in the paper. There is a growing need to address the concerns of local communities adjacent to protected areas, and ecotourism has the potential to provide many such benefits. People issues increasingly need to be integrated with parks issues and the fields of social ecology and anthropology, political economics and community-based natural resource management (CBNRM) can help build a bridge to connect the two. The 'Swi ta Lunga' Trust for example is an emerging Integrated Conservation and Development initiative in South Africa that attempts to integrate community development with ecotourism and conservation objectives.

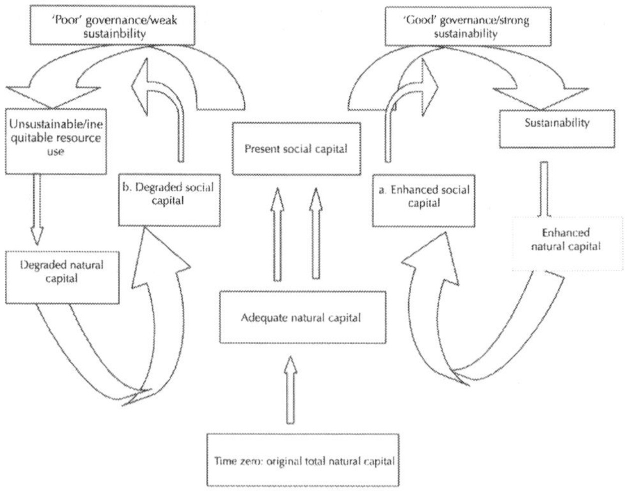

Figure 2: Natural capital providing the basis for social capital, which in turn provides the governance that determines the ongoing sustainability or deg-

radation of natural capital, which in turn drives enhanced (a) or degraded (b) social capital through desirable or undesirable feedbacks, respectively.

While some social scientists have been criticised for being too broad in their analysis and 'soft' in their approach, some natural scientists have been criticised for too easily dismissing the influence of societal components. Some social scientists argue that 'tunnel vision' may prevent such biophysical scientists of taking full account of human decisions, behaviour and relationships. These include issues such as power, equity, trust, ownership, participation, dignity, rights, justice, legitimacy, beliefs and attitudes, governance, cultural linkages and indigenous knowledge.

These cardinal aspects that impinge on the natural environment are as important to understanding and resolving issues as are biodiversity, biomass production, nutrient cycling and all the other fundamental themes in say mainstream ecology. We reiterate our contention that the social and ecological themes are interconnected, and those interconnections are often the missing links in the overall story we are trying to piece together. For this very reason, it becomes important to define and strengthen such mutual and integrated links in environmental management.

ECOSYSTEM SERVICES AND THE 'CAPITALS'

In this section, we present a simple model that aims to describe conceptually how natural capital differs from social capital, but how the two are inextricably interdependent. More specifically, it shows how 'good' vs 'poor' natural capital and 'positive' vs 'negative' social capital leads to 'strong' vs 'weak' sustainability depending on the effectiveness of the specific governance structure in place [Figure 2].

Conventional economics identifies three production factors, namely land, labour and capital (man-made fixed capital such as infrastructure). To biophysicists, this corresponds to natural (incorporating the essential livelihood ecosystem services described above), cultural/social and cultivated/manufactured capital.

Natural Capital: comprises the planet's non-renewable resources such as oil, minerals, etc., and renewable ones, comprising natural and mostly unmanaged ecosystems, which, importantly, maintain themselves (and their provision of essential outputs/services) at little or no cost to people. Natural capital is measured in terms of ecosystem services, biodiversity, species richness and general natural significance.

Cultural/Social Capital: includes the knowledge and traditions people use to make decisions, exploit resources (and each other), manufacture products and create values of our world. Social capital is measured in terms of employment created, GDP, returns on investment, but in general can include (more difficult to measure) softer social capital (e.g. trust in relationships).

Cultivated/manufactured capital: includes crops and forestry plantations and buildings, cars, etc. and derives from natural capital, although this fact is usually overlooked. Critically, unlike natural capital, cultivated and manufactured capital needs external energy sources and labour to produce and/or maintain them. Thus, they require significant resource input (investment of social and financial capital) before any output is realised, unlike natural capital.

Total capital: A capital asset is defined as a stock that yields a flow of goods and services. A society's total capital assets may therefore be broadly classified into manufactured, social capital and natural capital (non-renewable and renewable).

Natural, social and manufactured capital can be explained in the context of economic use values and in this regard conform to the principles of ecological economics as one of the primary effective integrating tools (see [Table 2]). Manufactured capital would represent the degree to which 'natural capital' is converted to 'social capital'. More specifically, social capital is applied to (renewable and non-renewable) natural capital to produce manufactured capital. There is a co-evolutionary interdependence between renewable natural capital and human social capital. Social capital adapts to the existing natural capital and to changes in it; natural capital changes as a result of the influence of social capital. Manufactured capital is operated by social capital for the extraction and transformation of natural capital to produce goods and services (Tacconi 2000).

Table 2: Values common to both natural and social sciences based on total economic value theory (after Tacconi 2000) and the Millennium Ecosystem Assessment (MA 2005)

	Use values		Non-use values
Direct use values	Indirect use values	Future use/option use values	
General utility value	Supporting services	Research value	Existence values
Natural capital Social capital Manufactured capital	Provisioning services Regulating services Cultural services	Bequest value	Aesthetic values Spiritual values Cultural values
			Peace values Wilderness values Permanence values

The issue of conversion of natural capital into manufactured or social capital is crucial, particularly the situation where some countries or operations deplete their natural capital without successfully creating sustainable manufactured or social capital in its place. This we term as 'poor' social governance, which includes not only poor attention to sound conservation (biophysical) principles, but the lack of equity in access to resources, appropriate legislation, intergenerational, etc. When this happens, the situation becomes increasingly less sustainable, creating an undesirable feedback loop. This is in contrast to the sustainable situation whereby social and manufactured capital create a desirable feedback loop of sustaining natural capital, enabling continual growth of appropriate social capital (see [Figure 2]).

We do not suggest by proposing this model that societies rich in natural capital will automatically and always generate 'good' social capital and vice versa, but rather both are needed to complement each other, to achieve sustainability (Stern 2004). Our model [Figure 2] recasts the common sustainability diagram with the three overlapping circles (environment, social and economic; see [Figure 1] in a way showing that while, in an evolutionary sense, natural capital is the basis, that in an ongoing management sense, social capital actually holds the key to maintaining sufficient functioning natural capital for generating sustainable livelihoods, through its 'good' or 'poor' governance decisions.

It is insufficient therefore to recognise only the need for taking account of both natural capital and social capital; it is how the two interconnect that is crucial, i.e. the linkages rather than the components. This we suggest is not a simple one-to-one relationship of two equal but competing parts being tradedoff against each other as so often happens; rather it is a complex system of each superimposed on the other ([Figure 2]; see also Gribbin 2005; Norgaard and Bear 2005).

By this superimposition we mean that a minimum amount of natural capital (incorporating provisioning services) must be present to underpin the existence of all and any social capital. However, this is essentially the 'Time Zero' situation (the notion of 'time' in this context to be viewed specifically in an evolutionary sense-see [Figure 2]); at all subsequent time periods, both natural and social capital are present, and it is today largely the social capital that determines the ongoing condition of the natural capital base and its ability to continue to deliver services.

Thus, there is a requirement for the 'right' (good) social capital to provide the political governance needed to ensure the continued existence of that adequate amount of natural capital, which, in turn, provides the continuous positive feedback needed for the ongoing growth of social capital. The 'wrong' (poor) social capital will of course lead to degradation of the natural capital resource base, leading to consequent degradation of the now unsupported social capital and ultimately collapse of both natural and social systems.

While we realise that we are casting a value judgement here as to 'right' and 'wrong', we are doing so in the belief that the one leads to sustainability and the other to non-sustainability. In this context, it can be useful to view different approaches to environmental management as opposite ends of a continuum. On the one extreme, the so-called

of the systems is increased rather than decreased by such stretching (Allison and Hobbs 2006). Those who advocate an exaggerated faith in human adaptability tend to represent a 'weak' sustainability framework and may not recognise natural limits and constraints. In this context, one needs to know how to 'deal with' such complexity (Norgaard and Bear 2005). Finally, humans are entirely dependent for their survival on at least a certain level of biophysical 'natural capital' as a resource base, a consideration which, we contend, needs recognition for social components to be successfully addressed.

As the Millennium Assessment (MA 2005) has shown, the global stock of natural capital has already been depleted to such an extent that it can no longer support the development aspirations of less-developed countries. For ecosystems and biodiversity to continue to provide life-supporting goods and services, a certain critical mass needs to be maintained. As we do not know exactly how much to provide for what levels of livelihoods for what numbers of people over what periods of time it may be wise to adopt the 'precautionary principle' and err on the conservative side.

The two case studies described next were chosen to depict the difference between 'good' and 'bad' natural capital, 'positive' and negative 'social capital', and 'good' and 'poor' governance, and how these factors influence the degree of potential sustainability (i.e. strong vs weak sustainability). More specifically, the ecotourism study (Case Study 1) shows that with 'good' governance we can enhance our natural capital in such a way that it continuously enhances social capital through socio-economics such as job creation, community ownership, etc. The IAS studies (Case Study 2) provide very clear examples of how, with 'poor' social governance, i.e. the deliberate or accidental introduction of IAS, immediate negative environmental impacts result, which in turn drive serious negative social impacts. (It should be noted that these examples could act in the reverse fashion if the governance are bad vs good, respectively). The point is that 'good' social governance and environmentally responsible behaviour would act in the reverse fashion and facilitate more sustainable solutions.

Case Study 1: The Potential of Ecotourism in South Africa-'good' natural capital catalysing 'positive' social capital leading to further investment in 'good' natural capital.

The International Ecotourism Society defines ecotourism as, 'responsible travel to natural areas that conserves the environment and improves the welfare of local people' (Brett 2005). From figures supplied by the World Tourism Organization and the IUCN, it is apparent that many of the leading tourist destinations throughout the world are also countries that maintain a network of protected areas. Africa attracts less than 2 percent of the international pool of long-haul tourists, which is an indication of a potential for significant growth. South Africa receives 1.7 million foreign (non-African) tourists a year, and eight countries, which account for only 18 percent of all arrivals generate 40 percent of tourism revenue. In 2003, the World Travel and Tourism Council estimated that the total tourism economy contributed 7.3 percent to GDP, and could increase to 8.8 percent in 10 years.

South African tourism has experienced steady growth over the past decade and is the only sector of the South African economy that has shown both a growth in employment and an increasing contribution to GDP. In other sectors of the economy, employment has actually decreased in spite of a positive contribution to GDP. Direct employment in the tourism industry in 2003 was estimated (2003) at 491,741 jobs, or 2.9 percent of total employment, but employment in the broader travel and tourism economy is estimated at 1.1 million, or 6.6 percent of total employment. Within a decade this figure could expand to 751,100 jobs, or 3.7 percent of the national total.

It is admittedly difficult to quantify the share of total tourism receipts that can be allocated to ecotourism. Surveys in South Africa indicate that 60 percent of foreign tourists are interested in nature-based tourism, while in Botswana and Kenya wildlife viewing comprises 50 percent and 70 percent, respectively, of gross tourist income. Domestic tourism in South Africa accounts for 67 percent of total receipts, and protected areas are becoming increasingly popular among local tourists. The Kruger Park now receives 1.3 million visitors, a four-fold increase since the1970s, and by March 2005, SANParks hosted 3.4 million visitors and sold 1.06 million bed nights. Provincial reserves receive approximately half the number of visitors received by SANParks, and private reserves host at least one million visitors, possibly many more.

If ecotourism is examined as a category of land use, then it outperforms agriculture in the economic sense. South Africa's national parks currently generate R115 per hectare per annum, which compares favourably with income from stock farming and is more impressive considering that Kruger and Kgalagadi Transfrontier Park comprise 78 percent of all national park estate. Some protected areas closer to major cities perform even better. Golden Gate National Park earns an estimated R950 per hectare, five times more than expected income from dairy farming, and yet it does not contain a single 'big five' species. The Park has also created more than 120 jobs, far more than would be possible from any other land use. Table Mountain National Park receives 4.2 million visitors per year, many of them repeat visits. Despite not having the 'big five', it is the most visited national park in Africa. The close proximity of a national park to a major city helps to maintain Cape Town's position as a major tourist destination. British Airways flights to Cape Town and Johannesburg are among the airline's top ten long-haul destinations, the others being American cities, Hong Kong and Dubai.

Although state-run protected areas can indeed be profitable, especially if they are located near major cities, the economic contribution made by private game reserves is even more impressive. Research in the Eastern Cape indicates that private game reserves are earning in excess of R1800 per hectare per annum, fifteen times more than they could from stock farming or hunting but creating 3.5 times as many jobs at 5.7 times the average salary compared to farming (Brett 2005). This would suggest that seven private game reserves in the Eastern Cape, encompassing some 50,000 hectares of dry thicket and thornveld, could be earning R95 million, a substantial amount considering that the country's entire mohair industry generates R200 million annually.

On a continent where the vast majority of people are poverty stricken, and where subsistence farming generates less than R50 per hectare per year, the potential for private reserves to generate significant income for local communities should be investigated further. Private lodges established in communal areas bordering national parks, with the benefits accruing to local communities would likely exceed revenues from any other land use. Local communities are already benefiting from lodges in the Makuleke region of Kruger National Park and in the Kgalagadi, but there is much potential for growth (see case study on the 'Swi ta Lunga' Trust below). One of the greatest advantages of ecotourism, especially in remote districts which are often underdeveloped and far from government services, is the potential to create permanent jobs. A direct job is created for every

1.5-1.8 beds in a rest camp within a national park in South Africa. In the Okavango, when compared to trophy hunting, ecotourism earns three times more but pays out twelve times as much in salaries to local people.

Without the attraction of wildlife destinations such as the Okavango, Etosha, Zambezi Valley and Kruger Park, southern African destinations would find it difficult to compete in a highly competitive global market. Many other countries, such as Spain and Portugal, are within easy reach of major markets, have sandy beaches, predictable weather, scenic mountains, stunning architecture, fine restaurants and vineyards. As a holiday to southern Africa requires at least a 10-hour flight from Europe, without the extensive protected-area network, southern Africa would lose the critical mass needed to attract tourists.

Although the region currently receives a very small slice of total world tourism receipts for niche market tourism-such as ecotourism-location is not as important as it is for mass tourism. Niche market tourists are more discerning and are overflying destinations such as West Africa and East Africa and are travelling the extra distance to southern Africa. African governments have often regarded protected areas as potential engines of economic growth. Simultaneously, increasing pressure on the national treasury for social spending has undermined the funding of much needed capital investment in protected areas. We believe that the interest shown in the transfrontier parks, and the potential to create many more protected areas in lightly populated regions such as the Karoo, can be translated into tangible economic benefits.

We thus propose that additional space created for ecotourism through the transfrontier initiatives, additional private reserves and the creation of new national parks, must be maximised through partnerships with recognised private sector operators and adjacent communities. By so doing, it may indeed be possible in certain circumstances to perfect the win-win solutions for the benefit of all stakeholders.

Case Study 2: IAS-'poor' social governance leading to degraded natural capital leading to degraded social capital [adapted from Mathews and Brand (2004)].

This selection of examples of IAS and their devastating effect on livelihoods in Africa exemplifies the issue of the complexity of natural systems and the impossibility of uncoupling socio-economics from the environment.

Who benefits and who pays? This question is increasingly being asked in countries vulnerable to 'development proposals' that have unstated or unquantified risks, including those posed by IAS. IAS are those invasive plant or animal species that become established in a new environment, then proliferate and spread in ways that are deleterious to biodiversity and/or human interests. Moreover, ostensibly favourable aid or trade options may have devastating consequences when IAS risks are not ascertained and mitigated for. And the hardest hit are invariably the poor. Africa already has too many examples of unwanted alien species that have imposed severe constraints on development potential, through impacts on biodiversity, the productive use of land, water security, fire, human health and many other usually unforeseen impacts. And things are rapidly getting worse, as we will show below.

The spread of IAS is now recognised as one of the greatest threats to the ecological and socio-economic well-being of the planet (see, for example the Convention on Biological Diversity; www.biodiv.org). It is creating complex and far-reaching challenges that threaten both the natural biological riches of the earth as well as the well-being of people everywhere. Economic costs alone run into hundreds of billions of US dollars per year and substantial efforts are required to rectify the situation. Locally, in the Cape Floral Kingdom, the invasion of alien tree species has decreased availability of water supplies to local communities, increased the frequency and intensity of fires and threatens native biodiversity, justifying government expenditures of over US$40 million per year in control programmes.

Examples of IAS that threaten biodiversity and have negative social and environmental costs include:

Lantana camara: Lantana is indigenous to South America but has been introduced as an ornamental plant in over 50 countries worldwide. It has invaded much of sub-Saharan Africa, forming dense thickets that displace natural indigenous plant communities and reduce agricultural productivity. The thickets disrupt the access of livestock to grazing and water, interfere with farming and forestry activities, and increase the intensity of fires.

By encroaching pastures, it reduces carrying capacities and productivity of agricultural land, encroaching as a weed in a variety of crops such as coffee, cotton, bananas and sugarcane. Furthermore, the entire plant is toxic and ingestion of the leaves and fruit can poison livestock. In some areas, lantana thickets provide a breeding ground for tsetse flies, propagating sleeping sickness disease. Little else can grow in these thickets because the plant is allelopathic, i.e. it releases chemicals into the soil which prevent other species from germinating. The lack of understory cover results in increased soil erosion, particularly on steep slopes. Impacts on local indigenous plant communities can be devastating.

Parthenium hysterophorus: Congress weed is an aggressive invader native to Mexico. Thought to have been introduced to Africa through imported wheat grain (including 'food aid' into Ethiopia!), it has invaded extensive tracts of subtropical areas in sub-Saharan Africa and South Africa. It invades disturbed land which has already been degraded, and also, being allelopathic, can quickly dominate pastures and croplands. It has such a devastating impact in Ethiopia that it has earned a local name meaning 'no crop'. It is unpalatable to livestock, so its invasion leads to grazing shortages. If it is mixed with fodder it taints the meat and milk. Parthenium also poses a health risk to humans and livestock, causing allergic reactions.

Eichhornia crassipes: The water hyacinth, another South American native, is considered the world's worst invasive aquatic weed. The plant grows rapidly, forming dense mats on the water surface up to a metre in thickness blocking waterways, obstructing fishing and other traffic. It impedes access to water by rural communities and their livestock, clogs intake pipes for water supply facilities, irrigation and hydroelectricity and has spread considerably on water bodies in South Africa. It also poses a health risk, providing an ideal habitat for mosquitoes and other disease vectors. By blanketing the surface it cuts out light and oxygen supply, creating conditions unsuitable for aquatic life, destroying fisheries in the process and affecting the quality of drinking water. Whole communities have been forced to abandon their villages as a result of infestation of local water bodies.

Lates niloticus: Although not a problem in South Africa itself, the Nile perch was introduced intentionally by British authorities into Lake Victoria to supplement the indigenous fish stocks, which were being severely depleted by overfishing as a result of an influx of migrant labour to work on nearby tea, coffee and sugar plantations established by these authorities. The indigenous fish fauna of the lake comprised over 400 species of small cichlid fish, making up 80 percent of biomass and about 20 non-cichlid genera.

With no natural predators and an abundance of prey, the Nile perch flourished. In less than 30 years, cichlids had declined to less than 1 percent of biomass, with over 200 species already extinct, and perch making up over 80 percent. But the bigger, oilier perch could not be sun dried like the smaller cichlids and required drying over fires, leading to devastation of woodlands, increasing erosion, increasing nutrient loads in the lake and sparking an infestation of water hyacinth.

The Nile perch also was in demand internationally, pushing prices up beyond the reach of locals and stimulating overfishing, the catch declining by over 80 percent by 2000. Not only has this introduction caused the single biggest mass extinction of vertebrates in recorded history, it has destroyed the sustainable indigenous fishery, destroyed whole communities and cultures based on those fisheries and today has left little of benefit in its place.

Mononychellus tanajoa: Some of the most damaging IAS in South Africa are agricultural pests, which can have crippling socio-economic impacts. While many introduced bacteria and viruses cause diseases of crops and livestock, insect pests are perhaps even more destructive. M. tanajoa, the cassava green mite is a major pest of cassava, a starchy root crop that is the staple food of more than 200 million people in sub-Saharan Africa. A native of South America, it was first detected in Africa in Uganda in the 1970s, thereafter spreading to more than twenty-five countries in Africa, including South Africa, causing an estimated 50 percent reduction in yield of the cassava crop, with all the attendant negative socio-economic impacts.

Prostephanus truncates: The larger grain borer is a destructive pest of farmstored maize and dried cassava in sub-Saharan Africa. Native to Central America, it was first detected in Africa in the late 1970s in Tanzania, where it increased maize losses by up to five times. It subsequently spread east, west and south, reaching South Africa in 1999. It is particularly damaging to grain stored on the cob after harvesting, but also stored cassava, cereals, legumes, tuber, peanuts, cocoa and coffee beans. To date, biological controls have been ineffective, requiring chemical control, with additional attendant negative human and environmental health risks.

NATIONAL BIODIVERSITY FRAMEWORK

The two case studies above represent examples of a range of environmentally related issues and challenges facing South Africa. South Africa's biodiversity 'stock' is estimated to have declined by some 20 percent in the past century alone and continues to decline relentlessly, while the economy has grown considerably in the same period (Biggs and Scholes 2004).

Therefore, an exciting (albeit demanding) aspect of recent South African legislation (Biodiversity Act 10 of 2004) sets the stage for the formulation of a new National Biodiversity Framework, where a number of biodiversity conservation issues can be addressed. For example, such a framework must provide for 'an integrated, coordinated and uniform approach to biodiversity management by organs of state in all spheres of government, non-governmental organisations, the private sector, local communities, other stakeholders and the public. It also has to be consistent with national environmental management principles and any relevant international agreements, identify priority areas for conservation action and the establishment of protected areas, and reflect regional co-operation on issues concerning biodiversity management in southern Africa.' (National Environmental Management Act: Biodiversity Act 10 of 2005; 39[1]).

The National Biodiversity Framework promotes the mainstreaming of biodiversity and requires an integrated approach in all sectors of society including agriculture, infrastructure, forestry, fisheries, energy, transport, water and sanitation. This framework, we suggest, can best be informed by two cross disciplinary fields: (1) environmental law and (2) ecological economics.

- Environmental law is indicative of the 'stick' approach and can be used in situations requiring prescriptive measures. This could include abandoning forestry policies that encourage resource degradation and the conversion of forest ecosystems to other less valuable uses, regulating the transfer of species and genetic resources and their release into the wild, asserting national sovereignty over genetic resources and regulating their collection, reforming policies that result in the degradation and the loss of biodiversity in freshwater, coastal and marine ecosystems; eliminating agricultural policies that promote excessive uniformity of crops or that encourage the overuse of chemical fertilisers and pesticides, i.e. environmental law could help to disincentivise and penalise counterproductive behaviour. The different drivers of global and of local agricultural production, pricing and marketing must become a central feature of national policy formulation.

- In contrast, ecological economics is indicative of the 'carrot' approach and can be used in situations that are more conducive to economic incentives. This might include the modification of national income accounts to make them reflect the economic losses that are incurred when biological resources are degraded and biodiversity is lost, a measure that would create an incentive to preserve such resources in the future. Other options include auditing the consumption of biological resources to raise awareness of the balance between local consumption and production thereby encouraging stakeholders to manage such resources more sustainably; the strengthening of institutional property rights over our natural resources; reforming policies that result in the degradation and the loss of biodiversity in coastal and marine ecosystems through privatisation of particular resources; establishing incentives for effective and equitable private-sector plant breeding and research; and reducing resource consumption through recycling and conservation with positive financial returns on environmentally responsible behaviour.

The National Biodiversity Framework can also benefit from other recent legislation such as the Protected Areas Act as setting aside protected areas play a key role in conserving biodiversity and maintaining natural capital. Such protected areas and how they relate to neighbouring communities is discussed next.

PEOPLE AND PARKS FOR PROSPERITY

The emerging paradigm of CBNRM is becoming a priority in environmental policy planning as southern Africa tackles human population growth and the pressure of impoverished communities on the boundaries of its reserves (see for example Barnes 1995; Brandon and Wells 1992; Chambers 1998; Magome and Fabricius 2004; DEAT 2003a,b;). In addition, 'People and Parks' issues are being addressed by various stakeholders in South Africa, such as the People and Conservation Department of South African National Parks, the Peace Parks Foundation and various research initiatives (such as TPARI-The 'Transboundary Protected Areas Research Initiative', which runs under the auspices of the IUCN). Although recently under challenge from some quarters (see overviews in, e.g. Brechin et al. 2002; Wilshusen et al. 2002), the general focus is on attempting to build capacity and address the needs of neighbouring communities adjacent to the protected areas in question, thereby turning a potential threat into an opportunity in a win-win partnership that is in the interests of all (Braak et al. 2004).

Such socio-economic considerations are becoming increasingly important in biodiversity conservation. Further, to accomplish the challenges set out by the National Biodiversity Framework, there may be several broad recommendations relating to CBNRM that could form part of such a strategy. For example, first there is a need to quantify and bring to the attention of policy makers the many economic and financial benefits of integrating rural development activities with the conservation of biological resources and diversity. Second, there is a need to develop the capacity of local communities and the means whereby they can participate in conservation efforts, so as to ensure more immediate and real benefits to local people. Possible conflicts as well as potential for cooperation between the various national economic activities should be identified, and 'mainstreaming' biodiversity into other sectors can have positive socio-economic spin-offs. Finally, relevant policies and legislation in other sectors should be examined for possible applications to the conservation of biodiversity and community involvement in such work.

In order for conservation to be successful in Africa, there specifically needs to be an integral and participatory relationship between ecotourism operations and the communities living next to protected areas. An interesting example of such an Integrated Conservation and Development Programme is the 'Swi ta Lunga' (Tsonga for 'things will come right') Community Development Trust based in Huntington, Limpopo Province, which borders the Greater Limpopo Transfrontier

Conservation Area and Sabi Sabi Private Game Reserve in South Africa (Reynolds 2004a,b,c). This is an example of a project that attempts to combine political economics with conservation and natural management. It also serves as a good case study of the positive potential of ecotourism in providing certain benefits to neighbouring adjacent communities.

The 'Swi ta Lunga' Trust focuses on reducing poverty through the renewal of social capital-in this case the traditional values that enable this marginalised community to once again become resource owners, managers and investors (see Brockington et al. 2006). Whereas in conventional markets, the producers or buyers are able to manage their activities by paying for them, such local communities often have no means to coordinate their activities. They remain 'dependents' of the state. This has been reinforced by the often empty promise of 'delivery' by the central government. The Government has now been asked-and a policy document has been presented to Cabinet-to provide 'child and investment rights' that form budgets managed by the Community Trust on behalf of all children and all 'equal owner' residents.

Child rights are maximised by buying food that is (mainly) locally grown and at locally determined prices. In this way, a significant new economic activity arises that is protected from global markets. Payments made to parents for food delivered to pre-, junior and senior schools are taxed to cover the school fees required. This transforms the 'community' into a competent partner for government, and one which is able to circulate public funds locally (the income multiplier here is four to five times). State salaries and other expenditure, in contrast, 'leaks' away to towns producing a far lower local multiplier of around 1.3. Once the system is in place, parents can bid for larger Child Rights grants per month in return for paying higher school fees. In this way, the government becomes an increasingly active investor through local citizens organised in Community Trusts, and people pay increasingly for social services (consumption) with community bursary and other financial devices to provide social security for all children and adults.

Improved grazing, irrigation and other investment initiatives belong to all in a new modernisation of traditional joint ownership of the resource base. Members receive equal 'annual use rights' over each resource, such as gardens, grazing and irrigation. These rights are tradeable, giving rise to resource prices and hence to ownership 'dividends'.

Huntington is thereby reinventing itself as a 'Community as Business' initiative able to partner government and business and other communities. The model promotes the greater circulation of money within the village so that Huntington can become economically self-sustainable. While Huntington gains the ability to invest cash and member labour to improve the natural and social resource base and 'within' community services, it will be able to take the longer view and promote women and child rights, provide training on sustainable grazing practices, community gardens, community banking and skill development in the hospitality and tourism industry.

The project needs to be viewed in the context of the Greater Limpopo Transfrontier Conservation Area. Huntington is but one of the villages in the 'Ntirhiswano' Forum, one of eight forums on the Western boundary of the Kruger National Park which comprises some 187 villages in total, reflecting the immense extent of this challenge (see Brockington et al. 2006). In addition, there are a number of villages residing within and around the Limpopo National Park in Mozambique on the Eastern boundary of the Kruger.

One positive result for Huntington could be an African village in transformation. Rural development tourism from neighbouring game lodges like Sabi Sabi should encourage the Shangaan people in the area to display their culture in an authentic and uncontrived manner, thereby encouraging a renewal of pride in their customs and history. This should hopefully also eventually lead to an interactive partnership with such tour operators and witness a 'community' such as Huntington becoming a represented party and role model in the context of other villages on the western border of the Greater Kruger Park Transfrontier Conservation Area. The Huntington 'model' starts from solid economic dynamics the local multiplier and the modernisation of traditional relations within communities and between members and resources so that 'Communities as Business' are able to participate in regional development opportunities, often as equal partners (see www.peoplesagenda.co.za).

The primary issue in the context of this paper is whether the natural capital in the protected areas adjacent to such communities can create or lead to the enhanced social capital required to uplift such communities from subsistence farming and poverty. In particular, can the benefits flowing from conservation and ecotourism in the region,

REFERENCES

1. Allison, H. and R. Hobbs. 2006. Science and Policy in Natural Resource Management: Understanding System complexity. Cambridge University Press, Cambridge, UK.
2. Aronson, J., Milton, S.J. and Blignaut, J. (eds) 2007. Restoring Natural Capital - Science, Business and Practice. Island Press, Washington, USA.
3. Arrow, K.J., P. Dasgupta and K.G. Maler. 2003. Evaluating projects and assessing sustainable development in imperfect economies. Environment and Resource Economics 26: 647-685.
4. Barnes, J.I. 1995. Economic analysis of community-based wildlife utilization initiatives in Botswana. Development Southern Africa 12: 783-803.
5. Berkes, F.J., J. Colding and C.Folke (eds.) .2003. Navigating the Socio-Ecological Systems. Building Resilience for Complexity and Change. Cambridge University Press, Cambridge, UK.
6. Biggs, R. et al. 2004. Nature Supporting People. The Southern African Millennium Ecosystem Assessment. Integrated Report. CSIR, South Africa.
7. Biggs, R and R. Scholes. 2004. Historical Changes in Natural Capital in South Africa: An Approach Using Changes in Biodiversity. Environmentek, CSIR, South Africa.
8. Braak, L., E. Mhlongo, J. Sturgeon and H. Magome. 2004. Policy recommendations on issues relating to neighbouring communities. KNP, Scientific services, www.parks sa.co.za/conservation/scientific_services/ss_KMIneighbour.html.
9. Brandon, K., K.H. Redford and S.E. Sanderson. (eds.). 1998. Parks in Peril. People, Politics and Protected Areas. The Nature Conservancy, Washington, DC.
10. Brandon, K.E. and M. Wells. 1992. Planning for people & parks: Design dilemmas. World Development 20(4): 557-570.
11. Brechin, S.R., Wilshusen, P.R., Fortwangeler, C.F. and P.C. West. 2002. Beyond the square wheel: Towards a more comprehensive understanding of biodiversity conservation as social and political process. Society and Natural Resources 15: 41-64.

12. Brett M. 2005. The Eco-Tourism Potential of South Africa. In: Vision XIII, Endangered Wildlife Trust. pp. 146-150. Future Publishing, South Africa.

13. Brockington, D., J. Igoe and K. Schmidt-Soltau. 2006. Conservation, human rights and poverty reduction. Conservation Biology 20(1): 250-252.

14. Chambers, R. 1998. Us and Them: Finding a New Paradigm for Professionals in Sustainable Development. In: Whose Reality Counts? Intermediate Technology Publications, London.

15. Costanza, R.1994. Three General Policies to Achieve Sustainability. In: Investing in Natural Capital: The Ecological Economics Approach to Sustainability (eds. A. Jansson, M. Hammer, C. Folke and R. Costanza.). Island Press, Washington, DC.

16. Costanza, R. and H. Daly. 1992. Natural capital and sustainable development. Conservation Biology 6: 37-46.

17. Curtin, C.G. 2002. Integration of science and community-based conservation in the Mexico/U.S. borderlands. Conservation Biology 16: 880-886.

18. DEAT. 2003a. People, Parks and Transformation in South Africa: A Century of Conservation, a Decade of Democracy. DEAT, Pretoria.

19. DEAT. 2003b. CBNRM-Guidelines for the Implementation of Community-Based Natural Resource Management in South Africa. DEAT, GTZ & DAI, Pretoria.

20. Gribbin, J. 2005. Deep Simplicity-Chaos, Complexity and the Emergence of Life. Penguin Books, Johannesburg, South Africa.

21. Heinberg, R. 2003. The Party's Over: Oil, War, and The Fate of Industrial Societies. New Society Publishers. 288pp.

22. Jansson, A., M. Hammer, C. Folke and R. Costanza (ed.). 1994. Investing in Natural Capital: The Ecological Economics Approach to Sustainability. Island Press, Washington, DC.

23. Lele, S. and B. Norgaard. 2005. Practicing interdisciplinarity. Bioscience 55: 967-975.

24. Mainka, S., J. McNeely and B. Jackson. 2005. Depend on Nature-Ecosystem Services supporting Human Livelihoods. IUCN Publication-see www.iucn.org/mdg5.

25. Magome, H. and C. Fabricius. 2004. Reconciling Biodiversity Conservation with Rural Development: The Holy Grail of CBNRM. In: Rights, Resources and Rural Development. Community-Based Natural Resource Management in Southern Africa. (eds. C. Fabricius, E. Koch, H. Magome and S. Turner) pp 93-111. Earthscan, London.

26. Mathews S. and K. Brand. 2004. Africa Invaded: The Growing Danger of Invasive Alien Species. Global Invasive Species Programme Secretariat, SANBI, Kirstenbosch, South Africa. 80pp.

27. Mascia, M. et al. 2003. Conservation and the Social Sciences. Conservation Biology 17(3): 649- 650.

28. Milton S., Aronson J. & Blignaut J. 2005. Unite to save Earth's natural capital. Quest 2 (1): 39- 41.

29. Norgaard, R.B. and P. Bear. 2005. Collectively seeing complex systems: The nature of the problem. Bioscience 55: 953-960.

30. Reynolds, N. 2004a. 'SWI TA LUNGA'-Members Handbook. The People's Agenda. Godobo Business Solutions, South Africa.

31. Reynolds, N. 2004b. Huntington Trust and Community Development Forum-Forming Working Partnerships between Community, Government and Business-the Emergence of an Enabled Community. Gobodo Business Solutions, South Africa.

32. Reynolds, N. 2004c. Community Grazing-A model for Huntington and neighbouring communities. Gobodo Business Solutions, South Africa.

33. Rietbergen-McCracken, J & Abaza, H. 2000. Environmental Valuation-a Worldwide Compendium of Case Studies. Earthscan, London.

34. Stern, D. 2004. The rise and fall of the environmental Kuznets curve. World Development 32: 1419-1439.

35. Tacconi, L. 2000. Biodiversity and Ecological Economics-Participation, Values and Resource Management. Earthscan, London.

36. Wilshusen, P.R., S.R. Brechin, C.F. Fortwangler and P.C. West. 2002. Reinventing a square wheel: Critique of a resurgent 'protection paradigm' in international biodiversity conservation. Society and Conservation 15: 17-40.

Chapter 5

Overcoming Barriers to Seedling Regeneration During Forest Restoration on Tropical Pasture Land and yhe Potential Value of Woody Weeds

Amelia T. Elgar[1], Kylie Freebody[1,2], Catherine L. Pohlman[3], Luke P. Shoo[4*] and Carla P. Catterall[1]

[1]School of Environment, Environmental Futures Research Institute, Griffith University, Brisbane, QLD, Australia

[2]Tablelands Community Revegetation Unit, Malanda, QLD, Australia

[3]Centre for Rainforest Studies, The School for Field Studies, Yungaburra, QLD, Australia

[4]School of Biological Sciences, The University of Queensland, St Lucia, QLD, Australia

ABSTRACT

Combating the legacy of deforestation on tropical biodiversity requires the conversion to forest of large areas of established pasture, where barriers to native plant regeneration include competition with pasture grasses and poor propagule supply (seed availability). In addition, initial woody plants that colonise pasture are often invasive, non-native species whose ecological roles and management in the context of forest regeneration are contested. In a restoration experiment at two 0.64 ha sites we quantified the response of native woody vegetation recruitment to (1) release from competition with introduced pasture grasses, and (2) local facilitation of frugivore-assisted seed dispersal provided by scattered woody plants and artificial bird perches. Herbicide pasture grass suppression during 20 months caused a significant but modest increase in density of native woody seedlings, together with abundant co-recruitment of the prominent non-native pioneer wild tobacco (*Solanum mauritianum*). Recruitment of native species was further enhanced by local structure in herbicide-treated areas, being consistently greater under live trees and dead non-native shrubs (herbicide-treated) than in open areas, and intermediate under bird perches. Native seedling recruitment comprised 28 species across 0.25 ha sampled but was dominated by two rainforest pioneers (*Homalanthus novoguineensis, Polyscias murrayi*). These early results are consistent with the expected increase in woody vegetation recruitment in response to release from competitive and dispersive barriers to rainforest regeneration. The findings highlight the need for a pragmatic consideration of the ecological roles of woody weeds and the potential roles of "new forests" more broadly in accelerating succession of humid tropical forest across large areas of retired agricultural land.

INTRODUCTION

Approximately half of the world's tropical biomes have been subjected to some form of clearing (Asner et al., 2009). One of the major drivers behind deforestation of tropical forests is clearing for agricultural practices (Achard et al., 2002). Consequently, large tracts of continuous rainforest have been converted to fragmented patches of remnant forest and secondary regrowth, situated amongst mosaics of agricultural land

and cattle pastures (Turner and Corlett, 1996). This process threatens global biodiversity, and causes increased global carbon emissions and changes in ecosystem functioning (Bradshaw et al., 2008). In some tropical forest landscapes, areas that were initially cleared for pasture and cattle grazing are eventually abandoned, due to declining productivity of pasture grasses, ongoing soil degradation, invasion of unpalatable grasses and changing socio-economic incentives (Hobbs and Cramer, 2007; Grau and Aide, 2008).

Forest recovery may subsequently occur in these retired tropical pastures, but several ecological factors act to reduce colonization by rainforest plants, potentially leaving the landscape in a state of arrested succession (Holl et al., 2000; Kanowski et al., 2009). Competition plays an important role in these dynamics, because a persistent cover of pasture grasses and herbs can limit forest regeneration following the removal of grazing livestock, by restricting micro-climatic conditions required for seed germination and the access of newly recruited woody seedlings to light, soil moisture or nutrients (Holl, 2002). Additionally, recovery of tropical forests is also often limited by a lack of propagule (seed) supply, because the seeds of many rainforest tree and shrub species have short-duration viability and are quickly exhausted from the soil seed bank during prolonged land use (Uhl, 1987; Holl et al., 2000). Furthermore, they are typically produced within fleshy fruits, so that their seed dispersal is mediated by frugivorous vertebrates that do not frequently visit open pasture (Da Silva et al., 1996; Wunderle, 1997).

Rapid reforestation over areas that are sufficiently large to be ecologically useful requires management interventions to reduce these barriers to regeneration. A variety of such interventions have been explored in recent years by an emerging cohort of restoration practitioners (Shoo and Catterall, 2013). For example, suppression of pasture grasses through various methods can in some circumstances promote reestablishment of woody vegetation (Shoo and Catterall, 2013). Additionally, the presence of isolated paddock trees potentially attracts seed dispersing animals which may move from forest to pasture and thereby facilitate seedling establishment (Guevara et al., 2004; Manning et al., 2006). Installation of artificial perches may similarly encourage seed rain, although subsequent seedling recruitment appears limited when pasture grasses are present (Holl, 1998; Shiels and Walker, 2003; Graham and Page, 2012). By contrast, canopy shade provided by live established paddock trees could competitively limit the growth of pasture grasses and herbs as well as potentially improving micro-climatic and soil conditions toward a more favorable environment for rainforest seedling recruitment (Rhoades et al., 1998; Manning et al., 2006). However, for these and other potential management interventions, there has been only limited systematic assessment of their effectiveness (Shoo and Catterall, 2013). A further dimension of management complexity occurs because non-native invasive plants are often the first and most abundant woody species recruited into retired pasture; these could act as recruitment facilitators for native forest seedlings by both attracting frugivores and shading the ground, but their treatment in restoration is controversial because of the possibility that they may also have negative effects on forest recruitment (D'Antonio and Meyerson, 2002; Kanowski et al., 2008; Davis et al., 2011).

Here we investigate methods for encouraging recruitment of rainforest seedlings in retired tropical pasture by removing or manipulating certain ecological barriers. We use a realistically scaled management experiment established in the Wet Tropics uplands of north eastern Australia to quantify the short term response of native woody seedling recruitment to: (1) herbicide-induced release from competition with introduced pasture grasses; and (2) local facilitation provided by elements of habitat structure, specifically scattered trees, shrubs, and artificial bird perches. We test whether rainforest seedling recruitment increases when competitive and dispersive barriers to regeneration are reduced. We also describe the extent and pattern of co-recruitment by non-native invasive woody plants, and discuss the findings in the context of current understanding of old field restoration in the tropics, with a particular emphasis on the contentious role of woody weeds in efforts to reinstate forest over large areas.

METHODS

Study Area

An experimental restoration project ("Kickstart Pasture Conversion Trials") was established in November 2011 in areas adjoining the Mt

Hypipamee/Upper Barron section of the Wet Tropics World Heritage Area, on the Atherton Tableland, north eastern Australia. The native vegetation is complex notophyll to mesophyll vine forest (Tracey, 1982). Landscape vegetation cover at the time of this study was a mosaic of remnant forest, substantial areas of livestock pasture (from which forest was cleared mainly in the first 5–6 decades of the twentieth century), and regrowth forest of varying ages. Pasture areas were mainly used for cattle grazing, and by the end of the century were dominated by non-native tropical grasses, such as signal grass (*Urochloa decumbens*) and pasture legumes, with guinea grass (*Megathyrsus maximus*) and setaria (*Setaria sphacelata* var. *anceps*) also widely established, together with a wide variety of other planted and invasive species.

So far, three kickstart pasture conversion sites have been established, each on red basaltic soil in an area of retired pasture which slopes steeply northwards to an east-flowing gully bordering a large area of conserved rainforest. In this paper we restrict most analyses to two sites from which the longest data time series is available (CloudlandE and CloudlandW; Figure S1), located on a single property (Cloudland: 17° 27′ 59″ S, 145° 32′ 28″ E, 875 m elevation), and separated by approximately 400 m which includes a strip (150 m wide) of restored rainforest that was planted in 2007. Limited shorter-term data are also presented from the third, later-established, site (Ringtail: 17° 27′ 59″ S, 145° 32′ 28″ E, 821 m elevation, some 3 km from Cloudland). The Cloudland property experienced cycles of partial clearing and regrowth since before the 1940s, and was partly cleared and grazed in the 1980s and 1990s, until 2005 when the entire property was destocked followed by ongoing exclusion of all grazing livestock. At the commencement of the study, pasture grass cover remained dense and was taller (often 0.5–1.0 m) than when grazed, and was variably intermixed with scattered colonising woody plants 2–8 m tall, comprising either native rainforest or non-native invasive species, together with occasional other small patches of low (<1.5 m tall) shrub or vine growth. Nearby weather stations indicate an average annual rainfall of 1467 mm, with 75% of the total falling between December and April (2001–2013, 2.6 km away at 031184 McKell Road Alert) and average monthly minimum and maximum temperatures of 14.7 and 25.4°C respectively (1961–1990, 16.5 km away at 031029 Herberton Mowbray Rd).

Design of Experimental Restoration Treatments

Each site contained two types of plot 8–200 m apart within the retired pasture, each plot being a square 80 m by 80 m (0.64 ha) with one side abutting the rainforest edge: (1) the "Works plot" within which all experimental management works took place; and (2) the "Control plot," a similar delineated area of retired pasture which remained untreated (Figures S1, S2). Management interventions in each Works plot were of two types: (1) grass and herb suppression; and (2) installation of bird-attracting structures. In addition, we also investigated (3) the influence of pre-existing trees and shrubs.

Grasses, herbs and other low-growing pasture-associated plants were suppressed with repeated herbicide applications, at a frequency which depended on weather conditions and observed grass or herb regrowth, as judged during regular site inspections (in general, grass and herb growth was slower during the cooler winter months). To reduce external influences, a 5.0 m wide buffer around the perimeter of the Works plot was also treated with herbicide. Herbicides were either glyphosate (which has a broad spectrum action on all types of plant) or the grass-selective Fusilade (fluazifop-p butyl) and Verdict (haloxyfop-R-methyl). The initial herbicide treatment in each Works plot aimed to achieve comprehensive glyphosate coverage of all ground-level plants, with no attempts to locate or protect small native forest seedlings, very few of which were apparent. Subsequent treatments were either spatially selective sprays with glyphosate, to spare any recruiting native rainforest seedlings, or more generalized sprays of grass-selective herbicide. Concentrations and application methods were those typically used by experienced restoration practitioners during establishment and maintenance of rainforest replanting projects in the region. Decisions about the type and delivery of herbicide (e.g., application timing; broad-spectrum vs. grass-selective chemicals; broadcast or localized delivery; use of vehicle-based high-pressure spray or backpack) were made independently and progressively for each site, depending on local topography, how the vegetation had recently developed, and the type of herbicide involved. These decisions were also guided by the underlying goal of longer term cost efficiency within a management context. Overall, during 20 months there were

7–10 herbicide treatments (varying in different parts of the plot) at CloudlandE, and seven at CloudlandW; and during the 9 months at Ringtail there were 6–7 sprays. Some treatments involved broadscale grass suppression, while others were more spatially localized, and targeted specific clumps of herbs, grasses, shrubs or scrambling vines. The nature of these treatments requires close tailoring of actions to suit each local combination of climate and ground-cover species.

Each bird attracting structure (henceforth a "perch") was constructed from an existing multi-branched *Alphitonia petriei* (a common early successional tree), cut at the base, with branches pruned back to a standard form (resulting height about 3–4 m, diameter 5–10 cm with 3–5 branches, embedded in the ground to a depth of 0.5 m). At the base of each sapling there was a water tray (surface 38 × 25 cm, depth 15 cm) which filled mainly from ambient rainfall with occasional hand supplementation, together with two logs (each about 1.0 m long and 20–25 cm diameter). Nine perches were installed in each Works plot in a regular grid of 20 m spacing, with three rows of three perches at each of 20, 40, and 60 m away from, and parallel to, the forest edge.

The pre-existing trees and shrubs in each plot were mapped if they were >2.0 m tall. These consisted of two types: (1) small native regrowth trees (henceforth termed "live trees"), and (2) non-native woody shrubs, all of which were treated with herbicide as part of the initial Works plot establishment (henceforth termed "dead clumps"). Live trees were mostly *Alphitonia petriei*, *Acacia celsa*,*Rhodamnia sessiliflora*, and *Xanthophyllum octandrum*. Care was taken during herbicide applications to avoid these trees. Dead clumps were almost all either lantana (*Lantana camara*) or wild tobacco (*Solanum mauritianum*), and were killed by stem treatment with Starane (fluroxypyr). A few small localized patches of very low growing native shrub (e.g., *Rubus queenslandicus*) or other small woody non-native plants were also killed with herbicide treatment during initial plot establishment.

Field Data Collection

Here we present data from seedling recruitment surveys conducted 20 months after the Works plots were established at the two Cloudland sites (i.e., in July 2013), and 9 months after the establishment of the Ringtail Works plot. Seedling recruitment was measured in two types of systematic search area: strips and circles. Strips were positioned regularly throughout both the Works and Control plots, and circles were placed around perches, live trees or dead clumps. In both types of search, all recruits of woody-stemmed trees or shrubs were counted and identified to species level if they: (1) were seedlings (i.e., excluding a few non-native shrubs that re-sprouted from rootstock); (2) were >10 cm in height; and (3) belonged to any species (native or non-native) which develops a woody stem that can typically exceed 2.5 cm diameter as individuals grow. The height of each stem was also measured, up to a maximum of 2.0 m (after which diameter was recorded).

Strips were transects 20 m by 2 m, positioned parallel to the forest edge. An arrangement of four strips laid end to end spanned the full width of the plot, and this was repeated at eight distances from the forest edge (5, 15, 25, 35, 45, 55, 65, 75 m), giving a total of 32 search strips in each plot (Figure S2). Circles had a radius of 2 m, which was placed around either the main stem of tree life-forms or a marked approximate center point within a multi-stemmed clump (the latter being typical for lantana). The number of circles surveyed in a plot depended on how many live trees, dead clumps and perches were present; in the Works plot at CloudlandE these numbers were 7 live trees, 0 dead clumps and 9 perches; compared with 5, 10 and 9 respectively at CloudlandW, and 11, 3, and 9 respectively at Ringtail.

Data Analyses

Counts of recruited seedlings in strips and circles (see above) were standardized to a density unit of stems per 100 m^2 to account for the difference in sampling area between search types (i.e., 40 m^2 for strips and 12.6 m^2 for circles). To examine the effect of grass suppression, mean recruit densities were compared between Works and Control plots, with strips as replicates but excluding all strips that intersected with live trees or dead shrubs, to remove the potential confounding effect of structure on recruitment (no strips intersected with circles around perches; Figure S2). Resulting sample sizes were 25 and 26 strips for Works and Control plots respectively at CloudlandE, 21 and 22 respectively at CloudlandW, and 23 and 6 respectively at Ringtail. To test the effect of structural features when acting in combination with grass suppression, we restricted data analysis to the herbicide-

habitat structures (GLM site, structure, interaction $\chi^2 = 496.61, 2716.65, 164.77$; $P < 0.001, 0.001, 0.001$; Figure 3A). At CloudlandE, mean seedling recruitment was 8.8 and 38.8 times greater under perches and live trees respectively, than in open areas were only grasses had been suppressed (Table 1). At CloudlandW, the effect of perches was greater and live trees less, engendering a similar response in seedling recruitment to the two structure types (17.4 and 18.1 times greater for perches and live trees, respectively, relative to open areas; Table 1). The presence of a third structure type, dead clumps, at CloudlandW induced a larger recruitment response than either perches or live trees (22.9 times more than open areas; Table 1).

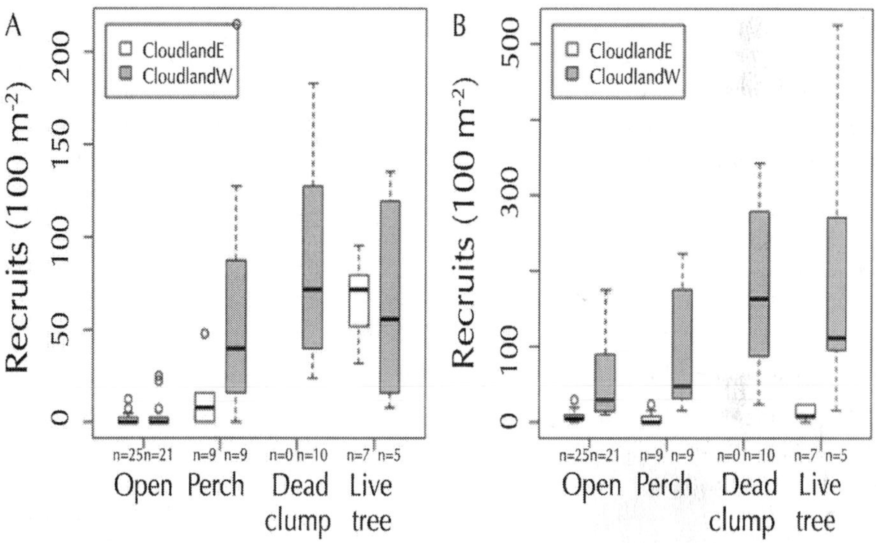

Figure 3: Effect of habitat structures (perches, dead clumps, live trees) on the density of recruited woody seedlings (>10 cm tall) after 20 months at two disused pasture sites (CloudlandE, CloudlandW): (A) native and (B) non-native species. Boxes encompass the upper and lower quartile of data with median value indicated by horizontal line. Whiskers are maximum or minimum values excluding outliers (circles) that are more than 1.5 times the upper or lower quartile.

The response of non-native seedlings was governed by a different form of interaction between site and structure (GLM site, structure, interaction $\chi^2 = 4345.5, 1294.4, 51.7$; $P < 0.001, 0.001, 0.001$; Figure 3B). At CloudlandE, mean recruitment of non-native seedlings was marginally less under perches (0.7 times) than in open areas but was enhanced by the presence of live trees (1.7 times greater; Table 1). The effect of perches and live trees was both positive and larger at CloudlandW, where seedling recruitment was 1.8 and 3.5 times greater under the two structure types respectively than in open areas. The effect of dead clumps was similar to that of live trees, with 3.1 times more recruitment than in open areas (Table 1), although the result for live trees was highly variable (Figure 3B).

Overall, the presence of structure also had a localized effect on the proportional representation between native and non-native seedlings. At CloudlandE, abundance of native seedlings was consistently greater than non-native seedlings under all structure types despite the opposite pattern in open areas where only grasses had been suppressed (Table 1). At CloudlandW, structure also increased proportional representation of native seedlings though non-native seedlings still remained more abundant (Table 1). The 9 month data for the Cloudland sites exhibited similar patterns in response to structure as did the 20 month data though the differences were less pronounced (data not shown). The third, younger, site (Ringtail, surveyed 9 months after establishment), showed a broadly similar pattern: herbicide alone caused a small increase in native and a larger increase in non-native recruitment, whereas live trees were associated with the largest increases in native recruitment (Ringtail site; Table 1).

Across both Cloudland Works plots, the mean height of native seedlings recorded in the analyzed strips and circles after 20 months was 61 cm (range 10–250 cm, median = 23 cm, with quartiles at 12 cm and 83 cm, $N = 338$; noting that minimum measurement threshold was 10 cm and individuals >2 m were assigned a height of 2.5 m).

For native recruits, the area-specific species richness values were strongly and positively correlated with their measured densities: for strips in open areas of Control and Works plots Spearman's $r = 0.99$ at CloudlandE and 0.99 at CloudlandW ($N = 51, 43$; $P < 0.001, 0.001$ respectively); and for circles across both types of plot Spearman's $r = 0.92$ at CloudlandE and 0.66 at CloudlandW ($N = 16, 24$; $P < 0.001, 0.001$ respectively).

DISCUSSION

Effects of Barrier-Lowering Interventions on Rainforest Seedling Recruitment

Almost no new seedlings emerged from the retired pasture Control plots over this study's 20 month period, supporting the idea that recruitment barriers are present. Repeated applications of herbicide to suppress the pasture grasses and herbs had a modest but discernable impact to enhance native seedling recruitment. This finding is broadly consistent with other regional and international studies where various interventions have been employed to alleviate ground conditions and promote the short term recovery of woody vegetation on degraded tropical land, although many of these did not differentiate between native and non-native species (Shoo and Catterall, 2013).

The recruitment of native rainforest seedlings in pasture where grasses had been suppressed was further enhanced by the presence of structural features, being greater under perches, live trees, and dead shrubs than in open areas. This is consistent with the notion that tropical forest trees can be both dispersal and establishment limited (Holl, 2012), and hence our findings add to a growing body of evidence that combined restoration interventions to simultaneously reduce multiple ecological barriers are more likely to yield tangible progress toward forest recovery than do individual techniques that target single ecological barriers (Holl et al., 2000; Posada et al., 2000; Ammondt et al., 2013). For example, in non-native grasslands of Hawaii, Brooks et al. (2009) found that broadcasting seed in conjunction with grass suppression (by herbicide application) was much more effective in enhancing the woody seedling recruitment than the summed effects of the individual techniques taken separately. Furthermore, strong correlations between species richness and abundance of natives reveal that the response of native rainforest seedlings is not just the result of one or two species but rather is underpinned by a suite of colonizing species.

Previous studies have demonstrated that perches can elevate seed rain into pasture but that this does not necessarily lead to increased seedling establishment (Holl, 1998; Shiels and Walker, 2003;Graham and Page, 2012). Our contrasting result likely stems from the fact that perches were installed in conjunction with active intervention to suppress competition from pasture grasses. However, the extent to which perches enhanced recruitment was modest when compared with the much greater recruitment beneath live trees or dead non-native shrubs. The importance of isolated trees as focal points for bird activity and seedling recruitment in pasture landscapes is well recognized (Toh et al., 1999; Manning et al., 2006). In our study, the larger effect of all trees and shrubs relative to perches is logical for two reasons. First, they had been established for a longer prior time-period, during which repeat frugivore visits could promote greater accumulation of deposited seeds. Second, the trees and shrubs had greater physical volume and architectural complexity, and potentially offered more food or shelter to attract visits by seed-dispersing birds. The dead non-native shrubs were present as a decaying skeletal framework during the 20 months of the study, and live native trees could have provided the birds with more food resources (fruit or insects) as well as shade to ameliorate local climate (thereby potentially increasing seedling germination or survival), but these factors appear relatively unimportant because the dead shrubs tended to produce greatest native seedling recruitment. This finding accords with the suggestion byToh et al. (1999) that the identity or fruiting status of trees matter less to the activity of seed dispersing birds near forest edges than their structure and suitability as perches.

We might also expect a spatial signature in the pattern of seedling recruitment because seed rain typically decreases with increasing distance from the forest edge, with most seed rain occurring closest to the forest edge (Willson and Crome, 1989; Cubiña and Aide, 2001). However, an exploratory analysis of the effect of distance from the forest edge did not reveal a discernible pattern in seedling recruitment, which instead was highly variable and inconsistent among sites. Given this finding, together with the limited sample size for structures, and also because distance from the edge could conceivably be correlated with several factors which may independently affect recruitment in different directions (seed rain, frugivore visits, slope, exposure to forest-associated marsupial herbivores), we did not include it as a factor in the analyses presented here.

Responses and Roles of Non-Native Woody Species

The most prominent short term outcome of herbicide-induced grass suppression in our study was the dramatic increase in the abundance and dominance of the non-native wild tobacco (*S. mauritianum*) which constituted the vast bulk of woody stems in open areas, and formed an extensive closed canopy 3–4 m tall over substantial parts of one experimental site (CloudlandW) within 20 months (Figure 4). Canopy dominance by a small number of light demanding tree species is not unusual in early stage succession (Guariguata and Ostertag, 2001). However, the non-native origin of the dominant species involved in this instance raises additional considerations for ongoing management.

Figure 4: Visual illustrations of different parts of a "Works" plot. Left panel—14 months after initial herbicide application (viewed from outside plot). Here, bird perches can be seen above the emerging canopy of non-native wild tobacco (*Solanum mauritianum*). **Right panel**—17 months after initial herbicide application (viewed from inside plot). Here, the cover of *S. mauritianum* has shaded the ground sufficiently to suppress pasture grasses.

Life history attributes of *S. mauritianum* include copious seed production and well-developed seed dormancy mechanisms, leading to high seedling recruitment from soil-stored seeds, and this species has been noted for its invasion potential (Florentine et al., 2003; Florentine and Westbrooke, 2003;Witkowski and Garner, 2008). Its relatively high soil seed storage explains why the ratio of recruit density in open areas vs. beneath habitat structures was higher for non-native woody recruits than was the case for native rainforest recruits (Table 1). Its fruit are fleshy and bird-dispersed, and hence seedling recruitment was nevertheless further enhanced by habitat structures.

This ability to colonise pasture, form a shady canopy, and quickly produce crops of fruit which are attractive to many seed dispersing frugivores (Florentine et al., 2003) may facilitate subsequent forest regeneration by greatly reducing ecological barriers of ground competition and propagule supply, much more rapidly and extensively than any native rainforest species appears capable of achieving. Indeed, Toh et al. (1999) suggested that lower strata trees such as *S. mauritianum* in the Australian subtropics likely perform an important function in providing the primary means by which taller canopy trees first enter former pasture sites. Our study's finding that native rainforest seedling recruitment was significantly enhanced beneath recently killed dead shrubs (among which *S. mauritianum* was well represented) supports the view that this species has ecological properties which facilitate forest succession. It is therefore possible that the new *S. mauritianum* stands which emerged after pasture suppression will provide positive feedback to enhance future seedling recruitment.

However, it is also possible that germination and growth of seedlings from the native seed rain falling into these stands will be suppressed by competitive interaction with the established *S. mauritianum*, necessitating future treatments to remove or thin the canopy in order to release a further rainforest seedling cohort. The notion that *S. mauritianum* may hinder recruitment of tropical rainforest pioneer and climax species stems partly from observations of an apparent absence of recruitment beneath the canopy of mature plants (Florentine et al., 2003). Shade house experiments show that concentrated aqueous leachates of *S. mauritianum* leaves inhibit the germination of lettuce seeds (*Lactuca sativa*) and can impair the early shoot and root growth of some native tropical rainforest trees (Florentine and Westbrooke, 2003). It is unknown whether these effects extrapolate to local field conditions and how they compare with the alleleopathic potential documented for some native pioneers (e.g., *Acacia melanoxylon*; González et al., 1995). In the Australian subtropics, low recruit abundances were reported beneath both *S. mauritianum* and the common native rainforest pioneer *Homalanthus nutans* which it resembles in stature (Toh et al., 1999).

The extensive recruitment of wild tobacco into retired pasture poses an important management dilemma, since it is necessary to weigh its potential benefits in reforestation against any undesirable ecological consequences of its dominance in early successional plant communities. In Australia, where suppression of invasive non-native species is a central aspect of contemporary restoration activities, routine control of *S. mauritianum* has been widely advocated in recent decades (e.g., Ward et al., 2001), although the species is not currently a declared weed under legislation and had been identified in an early ecological study as a transient successional species in forest regeneration (Williams et al., 1969). Clearly, there is a need for further research into the extent and mechanisms of the various competitive relationships that may occur between established trees and recruiting seedlings of both native and non-native woody species in an oldfield context.

More generally, there is an emerging viewpoint that while some non-native species can cause significant environmental problems, their removal may also have unforeseen consequences, and some may arguably be useful in restoration (D'Antonio and Meyerson, 2002; Reid et al., 2009). For example, in subtropical Australia, mature stands of the invasive tree *Cinnamomum camphora* in retired pasture support a high diversity of frugivorous birds that facilitate recruitment of a diverse suite of rainforest plants (Neilan et al., 2006; Kanowski et al., 2008). Particularly in moist tropical forest landscapes, some fleshy-fruited non-native tree species may thus provide opportunities for cost effective broad scale reforestation on retired agricultural land. In the restoration context, such realizations have prompted calls for non-native species to be judged on a broader understanding of factors such as their potential transience at a site and their roles in changing processes that influence the course of succession, rather than solely on their origins (D'Antonio and Meyerson, 2002; Davis et al., 2011).

Our findings likewise lead to a conclusion that evaluation of non-native pioneers' transience at sites is critical to making informed management decisions. Like many of its congeners, *Solanum mauritianum* is considered to be highly light-demanding and generally only persists and reproduces as long as the tree canopy remains relatively open (Murphy et al., 2008). This is consistent with evidence that the species declines over time during forest succession (Williams et al., 1969) and with declining light levels following closure of canopy gaps (Enright et al., 1993). In this study, many early-recruiting *S. mauritianum* individuals achieved the species' maximum height of about 4 m within less than 2 years. Most of the native species with which it co-recruited typically reach heights of at least 6 m (some considerably more), and so have the potential to overtop it. While species' growth rates vary, some *H. novoguineensis* individuals had grown as tall as, or taller than, the *S. mauritianum* within the 20 months of this study. Longer-term monitoring is needed to determine the capacity and time required for self-organized recruitment and growth of rainforest trees to shade out stands of *S. mauritianum*, and how well this development trajectory corresponds with time preferences for restoration outcomes.

CONCLUSIONS AND IMPLICATIONS

This study illustrates the role of shifting patterns of competitive inhibition, and its interactions with ecological facilitation, during old field succession in retired agricultural land. Initially pasture grasses outcompete rainforest seedlings, but any forest trees that do establish, as well as invasive woody shrubs, can then outcompete the grass. Subsequently there are also bidirectional potential competitive processes between native and invasive woody species. Understanding of the nature of interactions between different plant functional groups, in terms of both life-form and species origin, is fundamental to these processes, and has important practical applications in the imperative to achieve rapid reforestation of large areas of degraded agricultural land in the tropics. Strategic interventions can accelerate forest recovery, if appropriately selected and timed on the basis of this knowledge. Our findings reaffirm the importance of combined intervention approaches that are designed to overcome multiple ecological barriers, in some cases simultaneously and in others sequentially. Our findings also highlight the importance of retaining or actively reinstating "islands" of woody vegetation as regeneration foci in restoration (Zahawi et al., 2013).

Beneath habitat structures, especially live trees and dead shrubs, the recruited woody plant community at our study sites shifted substantially toward greater proportional representation of native seedlings. Thus, the presence of structure is likely to have important consequences for the longer term trajectory of community development.

A crucial gap in knowledge is the timeframe required for larger stature native tree species recruited into the seedling community to overtop the dominant non-native woody pasture invaders such as *S. mauritianum* and whether further interventions such as selective clearing or planting will be required to hasten this process. Our incidental field observations also suggest that the potential importance of herbivory by wildlife in mediating the recruitment of recruited seedlings in retired tropical pastures (see also Holl and Quiros-Nietzen, 1999) merits further investigation.

Empirical tests of the effectiveness of strategies other than active tree planting to simulate regeneration of tropical forest on degraded land have increased considerably over the last two decades (Shoo and Catterall, 2013). However, a major limitation to current knowledge is the short timeframes over which different interventions have been evaluated, and our study is no exception. Clearly, ongoing monitoring would be beneficial to reveal the longer-term trajectories of sites and the efficacy of alternative interventions employed to "kickstart" forest regeneration. In particular, such information would be instructive in determining whether the lower initial cost of less intense interventions considered here is offset by a longer and more expensive subsequent maintenance schedule than is typically needed after high density tree planting. Finally, this study has revealed a strong site based signature on rates of restoration progress, which means that experimental restoration projects need good landscape scale replication before the generality of findings can be fully ascertained.

ACKNOWLEDGMENTS

We thank Dave Hudson and Robyn Land (Cloudland Nature Refuge property) and Angela and Mark McCaffrey (Ringtail Crossing Nature Refuge property) for their interest in experimental restoration, for wide-ranging advice, and ongoing site access. Site works were conducted by Tablelands Regional Council Community Revegetation Unit with supervisory input from Larry Crook, and also involved Conservation Volunteers Australia (CVA, supervised by Alice Crabtree and Dave Hudson). Site works and some monitoring were funded by a grant from the Australian Government's Caring for Our Country program to the Wet Tropics Management Authority (WTMA) *"Mobilising landholders to improve landscape connectivity in the Wet Tropics,"* and by Toshiba via Landcare Australia. Thanks to Deb Pople, Campbell Clarke and Max Chappell of WTMA and Keith Smith (Qld Dept of Environment and Resource Management) for their input. Research and monitoring were part-funded by the National Environmental Research Program (Tropical Ecosystems and Environmental Decisions Hubs), by a WTMA student grant to Amelia T. Elgar, and were assisted by Amanda Freeman and students from The School for Field Studies. The investigation described here also formed part of Amelia T. Elgar's BSc (Honours) thesis at Griffith University (2013).

REFERENCES

1. Achard, F., Eva, H. D., Stibig, H.-J., Mayaux, P., Gallego, J., Richards, T., et al. (2002). Determination of deforestation rates of the world's humid tropical forests. *Science* 297, 999–1002. doi: 10.1126/science.1070656
2. Ammondt, S. A., Litton, C. M., Ellsworth, L. M., and Leary, J. K. (2013). Restoration of native plant communities in a Hawaiian dry lowland ecosystem dominated by the invasive grass *Megathyrsus maximus*. *Appl. Veg. Sci.* 16, 29–39. doi: 10.1111/j.1654-109X.2012.01208.x
3. Asner, G. P., Rudel, T. K., Aide, T. M., Defries, R., and Emerson, R. (2009). A contemporary assessment of change in humid tropical forests. *Conser. Biol.* 23, 1386–1395. doi: 10.1111/j.1523-1739.2009.01333.x
4. Bradshaw, C. J. A., Sodhi, N. S., and Brook, B. W. (2008). Tropical turmoil: a biodiversity tragedy in progress. *Front. Ecol. Environ.* 7, 79–87. doi: 10.1890/070193
5. Brooks, S., Cordell, S., and Perry, L. (2009). Broadcast seeding as a potential tool to reestablish native species in degraded dry forest ecosystems in Hawaii. *Ecol. Restor.* 27, 300–305. doi: 10.3368/er.27.3.300

6. Cubiña, A., and Aide, T. M. (2001). The effect of distance from forest edge on seed rain and soil seed bank in a tropical pasture. *Biotropica* 33, 260–267. doi: 10.1111/j.1744-7429.2001.tb00177.x

7. D'Antonio, C., and Meyerson, L. A. (2002). Exotic plant species as problems and solutions in ecological restoration: a synthesis. *Restor. Ecol.* 10, 703–713. doi: 10.1046/j.1526-100X.2002.01051.x

8. Da Silva, J. M. C., Uhl, C., and Murray, G. (1996). Plant succession, landscape management, and the ecology of frugivorous birds in abandoned Amazonian pastures. *Conserv. Biol.* 10, 491–503. doi: 10.1046/j.1523-1739.1996.10020491.x

9. Davis, M. A., Chew, M. K., Hobbs, R. J., Lugo, A. E., Ewel, J. J., Vermeij, G. J., et al. (2011). Don't judge species on their origins. *Nature* 474, 153–154. doi: 10.1038/474153a

10. Enright, N. J., Bartlett, R. M., and De Freitas, C. R. (1993). Patterns of species composition, recruitment, and growth within canopy gaps in two New Zealand kauri (Agathis australis) forests. *N.Z. J. Bot.* 31, 361–373. doi: 10.1080/0028825X.1993.10419514

11. Faraway, J. J. (2006). *Extending the Linear Model in R.* London: Chapman and Hall/CRC.

12. Florentine, S. K., and Westbrooke, M. E. (2003). Allelopathic potential of newly emerging weed *Solanum mauritianum*Scop. (Solanaceae) in the wet tropics of north-east Queensland. *Plant Prot. Q.* 18, 23–25. Available online at:http://www.weedinfo.com.au/ppq_abs18/ppq_18-1-23.html

13. Florentine, S., Westbrooke, M., and Craig, M. (2003). Flowering, fruiting, germination and seed despersal of the newly emerging weed *Solanum mauritianum* Scop. (Solanaceae) in the wet tropics of north Queensland. *Plant Prot. Q.* 18, 116–120. Available online at: http://www.weedinfo.com.au/ppq_abs18/ppq_18-3-116.html

14. González, L., Souto, X. C., and Reigosa, M. J. (1995). Allelopathic effects of Acacia melanoxylon R.Br. phyllodes during their decomposition. *For. Ecol. Manag.* 77, 53–63. doi: 10.1016/0378-1127(95)03581-T

15. Graham, L. L. B., and Page, S. E. (2012). Artificial bird perches for the regeneration of degraded tropical peat swamp forest: a

restoration tool with limited potential. *Restor. Ecol.* 20, 631–637. doi: 10.1111/j.1526-100X.2011.00805.x

16. Grau, H. R., and Aide, M. (2008). Globalization and land-use transitions in Latin America. *Ecol. Soc.* 13:16. Available online at: http://www.ecologyandsociety.org/vol13/iss2/art16/
17. Guariguata, M. R., and Ostertag, R., (2001). Neotropical secondary forest succession: changes in structural and functional characteristics. *Forest Ecol. Manag.* 148, 185–206. doi: 10.1016/S0378-1127(00)00535-1
18. Guevara, S., Laborde, J., and Sánchez-Rios, G. (2004). Rain forest regeneration beneath the canopy of fig trees isolated in pastures of Los Tuxtlas, Mexico. *Biotropica* 36, 99–108. doi: 10.1111/j.1744-7429.2004.tb00300.x
19. Hobbs, R. J., and Cramer, V. A. (2007). "Why Old Fields? Socioeconomic and ecological causes and consequnces of land abandonment," in *Old Fields: Dynamics and Restoration of Abandoned Farmland*, eds V. A. Cramer and R. J. Hobbs (Washington, DC: Island Press), 1–14.
20. Holl, K. D. (1998). Do perching structures elevate seed rain and seedling establishment in abandoned tropical pasture?*Restor. Ecol.*6, 253–261. doi: 10.1046/j.1526-100X.1998.00638.x
21. Holl, K. D. (2002). "Tropical moist forest," in *Handbook of Ecological Restoration*, Vol. 2. eds M. R. Perrow and A. J. Davy (Cambridge: Cambridge University Press), 539–558.
22. Holl, K. D. (2012). "Restoration of tropical forests," in *Restoration Ecology: The New Frontier, 2nd Edn*, eds J. Van Andel and J. Aronson (West Sussex: John Wiley and Sons, Ltd), 103–114. doi: 10.1002/9781118223130.ch9
23. Holl, K. D., Loik, M. E., Lin, E. H. V., and Samuels, I. A. (2000). Tropical montane forest restoration in Costa Rica: overcoming barriers to dispersal and establishment. *Restor. Ecol.* 8, 339–349. doi: 10.1046/j.1526-100x.2000.80049.x
24. Holl, K. D., and Quiros-Nietzen, E. (1999). The effect of rabbit herbivory on reforestation of abandoned pasture in southern Costa Rica. *Biol. Conserv.* 87, 391–395. doi: 10.1016/S0006-3207(98)00051-2

25. Kanowski, J., Catterall, C. P., and Neilan, W. (2008). Potential value of weedy regrowth for rainforest reforestation. *Ecol. Manage. Restor.* 9, 88–99. doi: 10.1111/j.1442-8903.2008.00399.x
26. Kanowski, J., Kooyman, R. M., and Catterall, C. P. (2009). "Dynamics and restoration of Australian tropical and subtropical rainforests," in *New Models for Ecosystem Dynamics and Restoration*, eds R. J. Hobbs and K. N. Suding (Washington, DC: Island Press), 206–220.
27. Manning, A. D., Fischer, J., and Lindenmayer, D. B. (2006). Scattered trees are keystone structures—Implications for conservation. *Biol. Conserv.* 132, 311–321. doi: 10.1016/j.biocon.2006.04.023
28. Murphy, H. T., Metcalfe, D. J., Bradford, M. G., Ford, A. F., Galway, K. E., Sydes, T. A., et al. (2008). Recruitment dynamics of invasive species in rainforest habitats following Cyclone Larry. *Austral Ecol.* 33, 495–502. doi: 10.1111/j.1442-9993.2008.01904.x
29. Neilan, W., Catterall, C. P., Kanowski, J., and McKenna, S. (2006). Do frugivorous birds assist rainforest succession in weed dominated oldfield regrowth of subtropical Australia? *Biol. Conserv.* 129, 393–407. doi: 10.1016/j.biocon.2005.11.007
30. Posada, J. M., Aide, T. M., and Cavelier, J. (2000). Cattle and weedy shrubs as restoration tools of tropical montane rainforest. *Restor. Ecol.* 8, 370–379. doi: 10.1046/j.1526-100x.2000.80052.x
31. R Core Team (2013). *R: A Language And Environment For Statistical Computing*. Vienna: R Foundation for Statistical Computing. Available online at: http://www.R-project.org/
32. Reid, A. M., Morin, L., Downey, P. O., French, K., and Virtue, J. G. (2009). Does invasive plant management aid the restoration of natural ecosystems? *Biol. Conser.* 142, 2342–2349. doi: 10.1016/j.biocon.2009.05.011
33. Rhoades, C. C., Eckert, G. E., and Coleman, D. C. (1998). Effect of pasture trees on soil nitrogen and organic matter: implications for tropical montane forest restoration. *Restor. Ecol.* 6, 262–270. doi: 10.1046/j.1526-100X.1998.00639.x
34. Shiels, A. B., and Walker, L. R. (2003). Bird perches increase forest seeds on Puerto Rican landslides. *Restor. Ecol.* 11, 457–465. doi: 10.1046/j.1526-100X.2003.rec0269.x

35. Shoo, L. P., and Catterall, C. P. (2013). Stimulating natural regeneration of tropical forest on degraded land: approaches, outcomes, and information gaps. *Restor. Ecol.* 21, 670–677. doi: 10.1111/rec.12048

36. Toh, I., Gillespie, M., and Lamb, D. (1999). The role of isolated trees in facilitating tree seedling recruitment at a degraded sub-tropical rainforest site. *Restor. Ecol.* 7, 288–297. doi: 10.1046/j.1526-100X.1999.72022.x

37. Tracey, J. G. (1982). *The Vegetation of the Humid Tropical Region of North Queensland*. Melbourne: CSIRO.

38. Turner, I. M., and Corlett, R., T. (1996). The conservation value of small, isolated fragments of lowland tropical rain forest.*Trends Ecol. Evol.* 11, 330–333. doi: 10.1016/0169-5347(96)10046-X

39. Uhl, C. (1987). Factors controlling succession following slash-and-burn agriculture in Amazonia. *J. Ecol.* 75, 377–407. doi: 10.2307/2260425

40. Ward, D., Goosem, S., andWerren, G. (2001). *Weed Pocket Guide: Agricultural and Environmental Weeds Far North Queensland*. Department of Natural Resources and Mines, Queensland Government. Available online at:http://www.wettropics.gov.au/site/user-assets/docs/FNQWeedBook.PDF

41. Williams, W. T., Lance, G. N., Webb, L. J., Tracey, J. G., and Dale, M. B. (1969). Studies in the numerical analysis of complex rain-forest communities: III. The analysis of successional data. *J. Ecol.* 57, 515–535. doi: 10.2307/2258396

42. Willson, M. F., and Crome, F. H. J. (1989). Patterns of seed rain at the edge of a tropical Queensland rain forest. *J. Trop.l Ecol.* 5, 301–308. doi: 10.1017/S0266467400003680

43. Witkowski, E. T. F., and Garner, R. D. (2008). Seed production, seed bank dynamics, resprouting and long-term response to clearing of the alien invasive Solanum mauritianum in a temperate to subtropical riparian ecosystem. *S. Afr. J. Bot.* 74, 476–484. doi: 10.1016/j.sajb.2008.01.173

44. Wunderle, J. M. Jr. (1997). The role of animal seed dispersal in accelerating native forest regeneration on degraded tropical lands. *Forest Ecol. Manag.* 99, 223–235. doi: 10.1016/S0378-1127(97)00208-9

45. Zahawi, R. A., Holl, K. D., Cole, R. J., and Reid, J. L. (2013). Testing applied nucleation as a strategy to facilitate tropical forest recovery. *J. Appl. Ecol.* 50, 88–96. doi: 10.1111/1365-2664.12014

Chapter 6

A Perspective on the Consequences for Insect Herbivores and Their Natural Enemies When They Share Plant Resources

Patrik Kehrli[1,2], and Steve D. Wratten[1]

[1]Bio-Protection Research Centre, Lincoln University, Liacoln 7647, Canterbury, New Zealand

[2]Département Protection des Végétaux Grandes Cultures et Vigne/Viticulture et Oenologie, Group Entomologie, Station de Recherche Agroscope Changins-Wädenswil ACW, CP 1012 1260 Nyon, Switzerland

ABSTRACT

Thousands of insect species consume both animal and plant-derived food resources. However, little recognition is given to the fact that omnivory is a general feeding strategy common to all higher trophic levels. Species in multitrophic interactions can all directly rely on the same plant resources. Nonetheless, little is known about the effect of a change in the relative abundance of a shared plant resource on trophic dynamics. Here we describe how a relative change of resource availability can affect multitrophic interactions and we emphasise its importance. Changes in multitrophic interactions can be induced by unequal alterations of individual fitness across trophic levels, possibly leading to changes in population structure of interacting species. At least ten ecological mechanisms can be involved and these are explored here. It is concluded that shared plant resources that are differentially used over several trophic levels have the potential to modify community structure and energy flow within food webs and ecosystems in more complex ways than previously recognised. The synthesis presented here provides an understanding of this complexity and can lead to improved deployment of biodiversity when manipulating food webs to protect ecological communities or to enhance ecosystem services such as biological control of agricultural pests.

INTRODUCTION

Herbivores are trapped between "the devil and the deep blue sea", that is, between natural enemies and plant defences [1–4]. However, little recognition is given to the fact that many natural enemies also directly rely on plant-derived food resources, such as nectar, pollen, seeds, or leaves [5–8]. Thousands of species of terrestrial arthropods such as predatory carabids, coccinellids, mites, spiders, and syrphids, along with hymenopteran and dipteran parasitoids, consume both animal and plant food; that is; they are omnivores [9, 10]. Omnivores can feed on animal-and plant-derived resources throughout their life, as can coccinellids, or utilise animal and plant food in different life stages, exhibiting life-history omnivory [11]. Examples are syrphids and parasitoids [9, 12]. Supplementing prey diet with plant food increases the individual fitness of omnivorous natural enemies [12, 13] and also sustains natural enemy populations when prey or hosts are scarce [11]. Hence, omnivory can promote top-down control [14]. Altogether, omnivores are important and persistent regulators of herbivore populations [15].

Omnivory can be found in species that are usually considered to be primary consumers (e.g., grasshoppers), in intermediate secondary consumers (e.g., predatory arthropods) and in top predators, such as bears or humans. Species in multitrophic interactions can therefore all directly rely on the same plant resources [14]. For example, floral nectar is frequently shared between butterflies, their parasitoids, and their hyperparasitoids (parasitoids whose hosts are other parasitoids) as a consequence of the spatial and temporal proximity of these organisms [16–18]. It is therefore common among arthropods that consumers in the herbivorous second (herbivores), the omnivorous third (natural enemies), and the omnivorous fourth trophic level (antagonists) all forage on shared plant resources (see Figure 1). However, little is known about the effect of a change in the relative abundance of a shared plant resource such as nectar on food web structure and dynamics. This perspective is to our knowledge the first synthesis of the mechanisms in which differences in the abundance or quality of shared plant resources can lead to asymmetrical fitness changes between and even within trophic levels. We describe how such a change of resource availability can affect trophic cascades, and we emphasise the importance of this for community ecology and biological control.

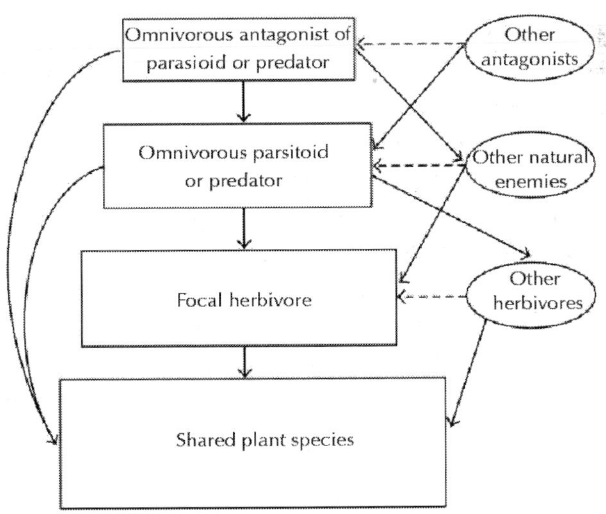

Figure 1: The main direct and indirect interactions among second trophic level herbivores, omnivores in the third trophic level, and omnivorous fourth

trophic level consumers foraging on a shared plant resource. Solid arrows indicate consumer-resource interactions whereas dotted ones indicate indirect interactions, such as interspecific competition. For simplicity, other interactions such as intraguild predation, apparent competition, and prey defensive responses are not illustrated. See text for detailed explanations.

HOW CAN SHARED PLANT RESOURCES ALTER TROPHIC CASCADES?

Trophic cascades are defined as reciprocal consumer-resource relationships that alter the abundance, biomass, or productivity of species' populations across several trophic levels of a food chain [19]. A change in the relative availability of a shared plant species can therefore not only affect the trophic interactions between herbivores, omnivorous natural enemies, and omnivorous antagonists, but also alter the trophic cascade within a subset of the food web community ("species-level cascades" sensu 2). Changes in species-level cascades are induced either by unequal alterations of individual fitness across the different trophic levels or by modification of the population structures of herbivores, natural enemies, and antagonists. At the level of individuals, relatively stronger fitness alterations at one level compared with the other two may result from different effects of changes in resource availability on the reproductive success of species [20–24]. At the population level, such changes can alter the sex ratio of offspring across trophic levels [25]. This is particularly important for parasitoids and hyperparasitoids because only females attack the herbivorous host. A higher proportion of parasitoid females in a population lead to increased parasitism rates, and this can lead to greater herbivore population suppression (see [26]). However, a higher proportion of hyperparasitoid females can lead to increased suppression of herbivores' natural enemies and consequently reduced top-down control of herbivores.

HOW ARE SPECIES-LEVEL TROPHIC CASCADES TRIGGERED?

A key question is how changes in relative availability of shared plant resources can have different effects on individual fitness or population structure across trophic levels. These effects can result from relative differences in the abundance, quality, or availability of shared plant resources, which can therefore favour one level of the trophic cascade over others by a range of mechanisms. These are summarised in Table 1 and are discussed in more detail here. (1) It is well known that the access of insects to floral nectar is determined by flower morphology [27]. (2) The attractiveness of plants, for example, shape, colour, and odour can differentially influence consumer guilds [27]. (3) Species in a trophic cascade are differentially affected by the chemical composition of plants, particularly the "secondary" compounds that determine the palatability of leaves, seeds, and pollen [28, 29]. Even easily accessible nectar differs in its sugar composition [30, 31], and Wäckers [32] showed that nectar consumers differ considerably in their demand for particular saccharides. Thus, even if a plant resource can be accessed by a number of consumer species, it may differentially affect them. (4) The temporal distribution of plant resource availability may favour one trophic level relative to the others because their phenology may only partially coincide [33]. (5) Mobility varies among species. Therefore, the spatial distribution of plant resources can differentially affect trophic levels [14, 34, 35]. (6) An asymmetrical enhancement of trophic levels within the food web may also result from the extent of species' reliance on shared resources [36]. The interaction strength between plants and their consumers is determined by consumers' feeding behaviour. For example, monophagous herbivores depend much more on a particular plant species than do omnivorous natural enemies. (7) Species-specific numerical and functional responses to the shared plant resource can favour members of one trophic level over the others [37, 38]. (8) Shared plant resources can modify a trophic cascade by altering species interactions with organisms outside the trophic cascade. For example, herbivores can suffer from increased interspecific competition [39, 40], and natural enemies can benefit from the presence of alternative prey [41] and can be exposed to altered intra- and interguild predation in the proximity of shared plant resources [42]. (9) Behavioural changes in species at different trophic levels can be induced by changes in relative availability of shared plant resources. For example, altered predation risk can induce behavioural changes in prey or hosts, such as predator avoidance behaviour and prey defensive responses [43]. These potentially costly behavioural changes may not only reduce prey/host reproduction but also feedback to higher-trophic-level consumers [44]. (10) Changes in relative availability of plant resources can directly affect the physiological condition of prey or hosts. This can occur as a consequence of altered intake of plant nutrients and toxins [45–47]. Altered prey or host quality can consequently affect higher-trophic-level consumers.

Table 1: A summary of how differences in the abundance, availability, or quality of a shared plant resource can lead to an asymmetrical enhancement of fitness across trophic levels. See text for references

(1)	Plant morphology
(2)	The attractiveness of plant resources
(3)	The nutritional value and chemical composition of plants
(4)	The temporal availability of plant resources
(5)	The spatial distribution of plants
(6)	The extent of species-specific reliance on the shared resource
(7)	Species-specific numerical and functional responses to the shared resource
(8)	Interactions with other organisms, such as competitors, pollinators, alternative prey, and intra- and interguild predators
(9)	Behavioural changes in species, such as predator avoidance behaviour prey defensive responses
(10)	Changed benefits to natural enemies and antagonists from prey/host consumption

In summary, changes in the relative availability of plant resources can indirectly amplify or mitigate top-down control [24, 45, 48]. This analysis of potential interactions within and between second, third, and fourth trophic levels mediated by a shared plant resource (Table 1) is certainly not complete. Even though most of these interactions are known to some extent, this is to our knowledge the first synthesis of the ways in which differences in the abundance, availability, or quality of shared plant resources can lead to asymmetrical enhancements of fitness within and between trophic levels. More research is needed to quantify the importance of each of these interactions for asymmetrical fitness enhancements within food webs. In the past, several studies modelled the influence of omnivory on multitrophic interactions and its persistence in predator-prey systems (e.g., [49–51]). In particular, Van Rijn et al. [14] developed a stage-structured predator-prey-resource model to explore the conditions under which individual plants gain protection from the plant-provided resources, such as extrafloral nectar and pollen, that are accessible to predators and herbivores. Their analysis revealed that the herbivore equilibria settle to lower values in the presence of supplementary food and that for the third trophic level the spatial distribution of shared resources is crucial. In conclusion, this synthesis of potential interactions within and between trophic levels (Table 1) provides a basis for understanding how trophic cascades can be triggered by shared food resources so that the implications of changes in the abundance of these resources can be predicted. This can in turn lead to more appropriate outcome-driven deployment of biodiversity in applied community ecology such as conservation and pest management. These aspects are discussed in more detail below.

IMPLICATIONS FOR COMMUNITY ECOLOGY

Omnivory is central to food web dynamics because it diffuses the effects of consumption across the trophic spectrum [52]. It increases web connectance and shortens food chains [53], as well as stabilising community interactions [54]. A strong change in the relative availability of a shared plant-derived resource, for example, as a result of plant invasion or extinction, may therefore alter the interactions among species (Figure 1) and as a result modify the entire food web [19]. A good example demonstrating the importance of plants in regulating the species composition of a community is the flower visitation web of a restored site in Mauritius [55]. Compared with a neighbouring, unrestored site that was heavily invaded by exotic plants, the restored site had a visitation web that was almost twice as species-rich and included visitors from several trophic levels. Also, the more complex pollinator community at the restored site allowed the native plant community to produce larger and heavier fruits with more seeds [55]. These results are supported by the broad literature on invasive and extinct plant species (e.g., [56–60]). Although invasive and endangered plant species are quite different in many aspects, both have the potential to play a disproportionately large role in shaping the community structure, especially if herbivores as well as omnivorous natural enemies and antagonists rely on these resources.

IMPLICATIONS FOR BIOLOGICAL CONTROL

About 36% of the world's food and fibre production is lost to plant pathogens, weeds, and insect pests (14%, 12%, and 10%, resp.) despite all types of pest control measures [61]. While chemical pesticide use can increase yields of crops, these compounds can be broad-spectrum toxins, which have external costs such as effects on human health, lethal accidents, damage to the environment, and disruption of ecosystem function [62]. Top-down control of pests by natural enemies is therefore an extremely important alternative or complement to chemical pest control [8]. The ecosystem service of biological control that is provided by natural enemies has been valued at >US$400 billion per year globally [63]. Thus, the manipulation of top-down control within agriculture is of large economic as well as biological interest. However, modern agricultural practices suppress biological control and functional biodiversity, so "ecological engineering" techniques are needed to enhance natural enemies' contribution to sustainable agriculture [64]. However, an overall enhancement of natural enemy biodiversity does not necessarily result in greater top-down control of pest populations [65]. As this diversity increases, so does the potential for interference between natural enemy species [66], intraguild predation [67], changes in the antipredator behaviour of prey and hosts [68, 69], and redundancy among natural enemy species [70]. In fact, recent evidence indicates that increased insect pest suppression may be best achieved by suites of natural enemies that forage in complementary spatial and temporal ways [70, 71]. The targeted deployment of specific plant resources that favour the fitness of selected biocontrol agents relative to their prey or host and their own antagonists can therefore boost top-down control [14, 72]. An asymmetrical enhancement of the fitness of targeted omnivorous predators and parasitoids can be achieved by using awareness of the factors and conditions given in Table 1. Implementing an asymmetrical enhancement of the effect of omnivorous biocontrol agents by the provision of alternative plant-derived food sources involves sophisticated mixtures of complementary crops in polycultures and intercropping, sophisticated weed management that preserves selective food sources, and the targeted cultivation of favourable noncrop species [7, 8, 64, 73, 74]. For example, targeted flowering plant species such as sweet alyssum, buckwheat, and sunflowers were intercropped with great success in Californian vineyards and vegetable fields. These plants increased the abundance of natural enemies such as hoverflies and parasitoids and consequently improved the biological control of aphids, leafhoppers, and thrips [75, 76]. These measures for providing alternative plant resources are economically attractive and help to conserve biodiversity and its associated ecosystem services, such as pollination, nutrient cycling, erosion management, and recreation. [8, 77, 78]. Overall, the selective enhancement of top-down control is one of the most promising tools in pest management in largely monocultural agroecosystems to ensure the conservation of vital ecosystem services for improved food and fibre production, without compromising environmental integrity and public health [79].

CONCLUSIONS

This novel perspective emphasises that relative differences in the abundance of plant resources not only affect invertebrate herbivores but also directly affect the fitness and abundance of omnivorous natural enemies that rely on such food sources. Plant resources that are differentially used over several trophic levels have the potential to modify community structure and energy flow within ecosystems in more complex ways than is currently recognised. It is important that ecologists and environmental managers are aware of this complexity when attempting to manage and manipulate food webs to protect ecological communities against invaders, maintain biodiversity, or enhance top-down control of agricultural pests. A better understanding on how trophic cascades are triggered by shared food resources will certainly enhance the success of nature conservation programs and pest management schemes.

ACKNOWLEDGMENTS

The authors thank Blas Lavandero, Don Vattala, Mauricio Urrútia, Salah Araj, Sam Scarratt, and Mark Wade for their useful suggestions, Matt Thomas, Teja Tscharntke, Bob Denno, Andrew Wilby, Bill Snyder, and Raphael Didham for discussing this research perspective, as well as Oliver Balmer, Jason Tylianakis, and Sue Zydenbos for their helpful comments on the manuscript. They gratefully acknowledge support from the Bio-Protection Research Centre, New Zealand. This study was funded by the Foundation for Research, Science and Technology, New Zealand (LINX 0303).

REFERENCES

1. S. B. Munch, M. L. Snover, G. M. Watters, and M. Mangel, "A unified treatment of top-down and bottom-up control of reproduction in populations," Ecology Letters, vol. 8, no. 7, pp. 691–695, 2005.
2. G. A. Polis, "Why are parts of the world green? Multiple factors control productivity and the distribution of biomass," Oikos, vol. 86, no. 1, pp. 3–15, 1999.
3. O. J. Schmitz, P. A. Hambäck, and A. P. Beckerman, "Trophic cascades in terrestrial systems: a review of the effects of carnivore removals on plants," American Naturalist, vol. 155, no. 2, pp. 141–153, 2000.
4. J. H. Lawton and S. McNeill, "Between the devil and the deep blue sea: on the problem of being a herbivore," in Population Dynamics, R. M. Anderson, B. D. Turner, and L. R. Taylor, Eds., Blackwell Scientific Publications, Oxford, UK, 1979.
5. F. L. Wäckers, P. C. J. van Rijn, and J. Bruin, Plant-Provided Food for Carnivorous Insects: A Protective Mutualism and Its Applications, Cambridge University Press, Cambridge, UK, 2005.
6. F. L. Wäckers, J. Romeis, and P. van Rijn, "Nectar and pollen feeding by insect herbivores and implications for multitrophic

interactions," Annual Review of Entomology, vol. 52, pp. 301–323, 2007.

7. D. A. Landis, S. D. Wratten, and G. M. Gurr, "Habitat management to conserve natural enemies of arthropod pests in agriculture," Annual Review of Entomology, vol. 45, pp. 175–201, 2000.
8. G. Zehnder, G. M. Gurr, S. Kühne, M. R. Wade, S. D. Wratten, and E. Wyss, "Arthropod pest management in organic crops," Annual Review of Entomology, vol. 52, pp. 57–80, 2007.
9. M. Coll and M. Guershon, "Omnivory in terrestrial arthropods: mixing plant and prey diets," Annual Review of Entomology, vol. 47, pp. 267–297, 2002.
10. W. E. Snyder and E. W. Evans, "Ecological effects of invasive arthropod generalist predators," Annual Review of Ecology, Evolution, and Systematics, vol. 37, pp. 95–122, 2006.
11. G. A. Polis and D. R. Strong, "Food web complexity and community dynamics," American Naturalist, vol. 147, no. 5, pp. 813–846, 1996.
12. M. D. Eubanks and J. D. Styrsky, "Effects of plant feeding on the performance of omnivorous "predators"," in Plant-Provided Food and Herbivore-Carnivore Interactions: A Protective Mutualism and Its Applications, F. L. Wäckers, P. C. J. van Rijn, and J. Bruin, Eds., Cambridge University Press, Cambridge, UK, 2005.
13. K. A. Robinson, M. Jonsson, S. D. Wratten, M. R. Wade, and H. L. Buckley, "Implications of floral resources for predation by an omnivorous lacewing," Basic and Applied Ecology, vol. 9, no. 2, pp. 172–181, 2008.
14. P. C. J. van Rijn, Y. M. van Houten, and M. W. Sabelis, "How plants benefit from providing food to predators even when it is also edible to herbivores," Ecology, vol. 83, no. 10, pp. 2664–2679, 2002.
15. M. D. Eubanks and R. F. Denno, "The ecological consequences of variation in plants and prey for an omnivorous insect," Ecology, vol. 80, no. 4, pp. 1253–1266, 1999.
16. L. R. Baggen, G. M. Gurr, and A. Meats, "Flowers in tri-trophic systems: mechanisms allowing selective exploitation by insect natural enemies for conservation biological control," Entomologia Experimentalis et Applicata, vol. 91, no. 1, pp. 155–161, 1999.

17. J. C. Lee and G. E. Heimpel, "Impact of flowering buckwheat on Lepidopteran cabbage pests and their parasitoids at two spatial scales," Biological Control, vol. 34, no. 3, pp. 290–301, 2005.

18. J. A. Rosenheim, T. E. Glik, R. E. Goeriz, and B. Rämert, "Linking a predator's foraging behavior with its effects on herbivore population suppression," Ecology, vol. 85, no. 12, pp. 3362–3372, 2004.

19. M. L. Pace, J. J. Cole, S. R. Carpenter, and J. F. Kitchell, "Trophic cascades revealed in diverse ecosystems," Trends in Ecology & Evolution, vol. 14, no. 12, pp. 483–488, 1999.

20. K. Winkler, F. L. Wäckers, A. Stingli, and J. C. van Lenteren, "Plutella xylostella (diamondback moth) and its parasitoid Diadegma semiclausum show different gustatory and longevity responses to a range of nectar and honeydew sugars," Entomologia Experimentalis et Applicata, vol. 115, no. 1, pp. 187–192, 2005.

21. B. Lavandero, S. D. Wratten, R. K. Didham, and G. Gurr, "Increasing floral diversity for selective enhancement of biological control agents: a double-edged sward?" Basic and Applied Ecology, vol. 7, no. 3, pp. 236–243, 2006.

22. S. E. Araj, S. Wratten, A. Lister, and H. Buckley, "Floral diversity, parasitoids and hyperparasitoids—a laboratory approach," Basic and Applied Ecology, vol. 9, no. 5, pp. 588–597, 2008.

23. P. Kehrli and S. Bacher, "Differential effects of flower feeding in an insect host-parasitoid system," Basic and Applied Ecology, vol. 9, no. 6, pp. 709–717, 2008.

24. M. Jonsson, S. D. Wratten, K. A. Robinson, and S. A. Sam, "The impact of floral resources and omnivory on a four trophic level food web," Bulletin of Entomological Research, vol. 99, no. 3, pp. 275–285, 2009.

25. L. A. Berndt and S. D. Wratten, "Effects of alyssum flowers on the longevity, fecundity, and sex ratio of the leafroller parasitoid Dolichogenidea tasmanica," Biological Control, vol. 32, no. 1, pp. 65–69, 2005.

26. D. L. J. Quicke, Parasitic Wasps, Chapman & Hall, London, UK, 1997.

27. H. F. Howe and L. C. Westley, "Ecology of pollination and seed dispersal," in Plant Ecology, M. J. Crawley, Ed., Blackwell Science, Oxford, UK, 1997.

28. J. P. Rosenthal and M. R. Berenbaum, Herbivores: Their Interactions with Secondary Plant Metabolites, Academic Press, San Diego, Calif, USA, 1991.
29. P. J. Ode, "Plant chemistry and natural enemy fitness: effects on herbivore and natural enemy interactions," Annual Review of Entomology, vol. 51, pp. 163–185, 2006.
30. H. G. Baker and I. Baker, "Floral nectar sugar constituents in relation to pollinator type," in Handbook of Experimental Pollination Biology, C. E. Jones and R. J. Little, Eds., Scientific and Academic Editions, New York, NY, USA, 1983.
31. H. D. Vattala, S. D. Wratten, C. B. Phillips, and F. L. Wäckers, "The influence of flower morphology and nectar quality on the longevity of a parasitoid biological control agent," Biological Control, vol. 39, no. 2, pp. 179–185, 2006.
32. F. L. Wäckers, "Gustatory response by the hymenopteran parasitoid Cotesia glomerata to a range of nectar and honeydew sugars," Journal of Chemical Ecology, vol. 25, no. 12, pp. 2863–2877, 1999.
33. C. A. Edwards, K. D. Sunderland, and K. S. George, "Studies on polyphagous predators of cereal aphids," Journal of Applied Ecology, vol. 16, pp. 811–823, 1979.
34. J. M. Bullock, R. E. Kenward, and R. S. Hails, Dispersal Ecology, Blackwell Publishing, Oxford, UK, 2002.
35. S. D. Wratten, M. H. Bowie, J. M. Hickman, A. M. Evans, J. R. Sedcole, and J. M. Tylianakis, "Field boundaries as barriers to movement of hover flies (Diptera: Syrphidae) in cultivated land," Oecologia, vol. 134, no. 4, pp. 605–611, 2003.
36. R. T. Paine, "Food-web analysis through field measurement of per capita interaction strength," Nature, vol. 355, no. 6355, pp. 73–75, 1992.
37. O. Gilg, B. Sittler, B. Sabard et al., "Functional and numerical responses of four lemming predators in high arctic Greenland," Oikos, vol. 113, no. 2, pp. 193–216, 2006.
38. S. M. Redpath and S. J. Thirgood, "Numerical and functional responses in generalist predators: hen harriers and peregrines on Scottish grouse moors," Journal of Animal Ecology, vol. 68, no. 5, pp. 879–892, 1999.

39. R. F. Denno, M. S. McClure, and J. R. Ott, "Interspecific interactions in phytophagous insects: competition reexamined and resurrected," Annual Review of Entomology, vol. 40, pp. 297–331, 1995.
40. A. J. A. Stewart, "Interspecific competition reinstated as an important force structuring insect herbivore communities," Trends in Ecology & Evolution, vol. 11, pp. 233–234, 1996.
41. R. J. Morris, O. T. Lewis, and H. C. J. Godfray, "Experimental evidence for apparent competition in a tropical forest food web," Nature, vol. 428, no. 6980, pp. 310–313, 2004.
42. D. L. Finke and R. F. Denno, "Spatial refuge from intraguild predation: implications for prey suppression and trophic cascades," Oecologia, vol. 149, no. 2, pp. 265–275, 2006.
43. E. H. Nelson, C. E. Matthews, and J. A. Rosenheim, "Predators reduce prey population growth by inducing changes in prey behavior," Ecology, vol. 85, no. 7, pp. 1853–1858, 2004.
44. E. H. Nelson, "Predator avoidance behavior in the pea aphid: costs, frequency, and population consequences," Oecologia, vol. 151, no. 1, pp. 22–32, 2007.
45. M. A. Urrutia C, M. R. Wade, C. B. Phillips, and S. D. Wratten, "Influence of host diet on parasitoid fitness: unravelling the complexity of a temperate pastoral agroecosystem," Entomologia Experimentalis et Applicata, vol. 123, no. 1, pp. 63–71, 2007.
46. H. C. J. Godfray, Parasitoids: Behavioural and Evolutionary Ecology, Princeton University Press, Princeton, NJ, USA, 1993.
47. R. Senrayan and R. S. Annadurai, "Influence of host's food plant and habitat on Anastrus ramakrshnae(Mani) (Hym., Eupelmidae), an egg parasitoid of Coridius obscurus (Fab.) (Het., Pentatomidae),"Journal of Applied Entomology, vol. 112, pp. 237–243, 1991.
48. J. A. Harvey, N. M. van Dam, and R. Gols, "Interactions over four trophic levels: foodplant quality affects development of a hyperparasitoid as mediated through a herbivore and its primary parasitoid,"Journal of Animal Ecology, vol. 72, no. 3, pp. 520–531, 2003.
49. M. van Baalen, V. Křivan, P. C. J. van Rijn, and M. W. Sabelis, "Alternative food, switching predators, and the persistence of

predator-prey systems," American Naturalist, vol. 157, no. 5, pp. 512–524, 2001.

50. V. Křivan and S. Diehl, "Adaptive omnivory and species coexistence in tri-trophic food webs,"Theoretical Population Biology, vol. 67, no. 2, pp. 85–99, 2005.
51. J. Kean, S. Wratten, J. Tylianakis, and N. Barlow, "The population consequences of natural enemy enhancement, and implications for conservation biological control," Ecology Letters, vol. 6, no. 7, pp. 604–612, 2003.
52. D. Simberloff and P. Stiling, "How risky is biological control?" Ecology, vol. 77, no. 7, pp. 1965–1974, 1996.
53. W. G. Sprules and J. E. Bowerman, "Omnivory and food chain length in zooplankton food webs,"Ecology, vol. 69, no. 2, pp. 418–426, 1988.
54. W. F. Fagan, "Omnivory as a stabilizing feature of natural communities," American Naturalist, vol. 150, no. 5, pp. 554–567, 1997.
55. C. N. Kaiser, Functional integrity of plant-pollinator communities in restored habitats in Mauritius, Dissertation, Universität Zürich, 2006.
56. J. Belnap, S. L. Phillips, S. K. Sherrod, and A. Moldenke, "Soil biota can change after exotic plant invasion: does this affect ecosystem processes?" Ecology, vol. 86, no. 11, pp. 3007–3017, 2005.
57. B. J. Brown, R. J. Mitchell, and S. A. Graham, "Competition for pollination between an invasive species (purple loosestrife) and a native congener," Ecology, vol. 83, no. 8, pp. 2328–2336, 2002.
58. J. E. Duffy, "Biodiversity and ecosystem function: the consumer connection," Oikos, vol. 99, no. 2, pp. 201–219, 2002.
59. W. H. van der Putten, P. C. De Ruiter, T. M. Bezemer, J. A. Harvey, M. Wassen, and V. Wolters, "Trophic interactions in a changing world," Basic and Applied Ecology, vol. 5, no. 6, pp. 487–494, 2004.
60. I. Bartomeus, M. Vilà, and L. Santamaría, "Contrasting effects of invasive plants in plant-pollinator networks," Oecologia, vol. 155, no. 4, pp. 761–770, 2008.

61. G. N. Agrios, Plant Pathology, Elsevier Academic Press, Amsterdam, The Netherlands, 5th edition, 2005.

62. G. J. Devine and M. J. Furlong, "Insecticide use: contexts and ecological consequences," Agriculture and Human Values, vol. 24, no. 3, pp. 281–306, 2007.

63. R. Costanza, R. d›Arge, R. de Groot et al., "The value of the world›s ecosystem services and natural capital," Nature, vol. 387, no. 6630, pp. 253–260, 1997.

64. G. M. Gurr, S. D. Wratten, and M. A. Altieri, Ecological Engineering for Pest Managment Advances in Habitat Manipulation for Arthropods, CSIRO Publishing, Collingwood, Australia, 2004.

65. B. J. Cardinale, D. S. Srivastava, J. E. Duffy et al., "Effects of biodiversity on the functioning of trophic groups and ecosystems," Nature, vol. 443, no. 7114, pp. 989–992, 2006.

66. M. D. Moran and L. E. Hurd, "Short-term responses to elevated predator densities: noncompetitive intraguild interactions and behavior," Oecologia, vol. 98, no. 3-4, pp. 269–273, 1994.

67. D. L. Finke and R. F. Denno, "Predator diversity dampens trophic cascades," Nature, vol. 429, no. 6990, pp. 407–410, 2004.

68. O. J. Schmitz, V. Krivan, and O. Ovadia, "Trophic cascades: the primacy of trait-mediated indirect interactions," Ecology Letters, vol. 7, no. 2, pp. 153–163, 2004.

69. R. P. Prasad and W. E. Snyder, "Diverse trait-mediated interactions in a multi-predator, multi-prey community," Ecology, vol. 87, no. 5, pp. 1131–1137, 2006.

70. P. Casula, A. Wilby, and M. B. Thomas, "Understanding biodiversity effects on prey in multi-enemy systems," Ecology Letters, vol. 9, no. 9, pp. 995–1004, 2006.

71. A. R. Ives, B. J. Cardinale, and W. E. Snyder, "A synthesis of subdisciplines: predator-prey interactions, and biodiversity and ecosystem functioning," Ecology Letters, vol. 8, no. 1, pp. 102–116, 2005.

72. J. M. Tylianakis, R. K. Didham, and S. D. Wratten, "Improved fitness of aphid parasitoids receiving resource subsidies," Ecology, vol. 85, no. 3, pp. 658–666, 2004.

73. D. A. Andow, "Vegetational diversity and arthropod population response," Annual Review of Entomology, vol. 36, no. 1, pp. 561–586, 1991.

74. T. Tscharntke, A. M. Klein, A. Kruess, I. Steffan-Dewenter, and C. Thies, "Landscape perspectives on agricultural intensification and biodiversity—ecosystem service management," Ecology Letters, vol. 8, no. 8, pp. 857–874, 2005.

75. R. Colfer, "Using habitat management to improve biological control on commercial organic farms California," in California Conference on Biological Control IV, M. S. Hoddle, Ed., pp. 55–62, Center for Biological Control, College of Natural Resources, University of California, Berkley, Calif, USA, July 2004.

76. C. I. Nicholls, M. P. Parrella, and M. A. Altieri, "Reducing the abundance of leafhoppers and thrips in a northern California organic vineyard through maintenance of full season floral diversity with summer cover crops," Agricultural and Forest Entomology, vol. 2, no. 2, pp. 107–113, 2000.

77. M. Loreau, S. Naeem, P. Inchausti et al., "Ecology: biodiversity and ecosystem functioning: current knowledge and future challenges," Science, vol. 294, no. 5543, pp. 804–808, 2001.

78. G. C. Daily and P. A. Matson, "Ecosystem services: from theory to implementation," Proceedings of the National Academy of Sciences of the United States of America, vol. 105, no. 28, pp. 9455–9456, 2008.

79. D. Tilman, K. G. Cassman, P. A. Matson, R. Naylor, and S. Polasky, "Agricultural sustainability and intensive production practices," Nature, vol. 418, no. 6898, pp. 671–677, 2002.

Chapter 7

From Planning to Implementation: Explaining Connections Between Adaptive Management and Population Models

José J. Lahoz-Monfort, Gurutzeta, Guillera-Arroita, and Cindy E. Hauser

Quantitative and Applied Ecology Group, School of Botany, University of Melbourne, Parkville, VIC, Australia

ABSTRACT

The management of natural systems often involves periodic interventions that must be decided without a complete understanding of how the system responds to our actions. It is in this situation of recurrent decision-making under uncertainty that adaptive management (AM) has been repeatedly advocated, with each decision round providing an opportunity to improve our knowledge in order to facilitate future decisions: the "learning while managing" tenet of AM. When the subject of management is a wildlife population (that is harvested, is a pest or is threatened with extinction), population models will be at the core of the AM process. We provide an overview of the steps in AM, from the set-up to the iterative phase, highlighting the central role that population models can play at different stages of the process of planning and implementing an AM program, as well as when analyzing the value of acquiring new information. We discuss the contexts in which these models have been applied in natural resource management and biodiversity conservation. We aim to bring this applied discipline to the attention of researchers interested in population dynamics, while stressing the relevance of these models for managers considering an AM approach.

INTRODUCTION

Management decisions regarding biodiversity conservation and natural resource management (NRM) often have to be made under uncertainty, which may arise from different sources (Regan et al., 2002; Williams et al., 2007, p. 61). When the management is directed at some animal or plant population rather than at an ecosystem or habitat, the uncertainties that impede straightforward decision-making are often underpinned by some aspect of population dynamics. From the specification of sustainable harvest rates (Nichols et al., 1995) to the management of pests (Shea et al., 2002) and endangered species (Runge, 2011), decisions need to be based on an understanding of population growth or demography, knowledge that can be formalized as population models. By population models we refer both to growth models (e.g., logistic or Ricker model; Ricker, 1954) and to models with explicit demographic dynamics (process models, e.g., based on a Leslie matrix; Caswell, 2001). The structure of such models, the value of the parameters within them, or how these are affected by environmental or human-related factors, are potential sources of uncertainty. Improving our knowledge about natural systems can be costly and/or time-consuming and waiting until uncertainties are resolved may not always be feasible in the face of pressing management decisions. In this context of recurrent decision-making under uncertainty "adaptive management" ("AM" hereafter) has often been advocated, such that knowledge can be updated over time and management actions adapted accordingly. It is therefore natural that population models play a key role in the application of the AM framework to programs that deal with populations of animals and plants.

The theory of AM was developed several decades ago in fisheries science (Walters and Hilborn, 1976, 1978; Smith and Walters, 1981) and then spread to other areas of NRM (Parma et al., 1998; Williams et al., 2007), particularly waterfowl hunting management (Nichols et al., 1995; Johnson et al., 1997). The last decade has seen a renewed interest in its application to biodiversity conservation, for both plant (Moore et al., 2011) and animal species (Rout et al., 2009; McDonald-Madden et al., 2010; Runge, 2013), as well as habitat-based approaches (Gerber et al., 2005; Moore and Conroy, 2006; McCarthy and Possingham, 2007). Although often dealing with a single species, multi-species problems have also been addressed using AM, including predator-prey systems with linked dynamics (Varley and Boyce, 2006; Martin et al., 2010) and sympatric species (Sainsbury, 1991; Smith et al., 2013).

In this paper we provide an overview of AM, stressing the central role that population models play in the different stages of the process of planning and implementing an AM program (Figure 1), as well as in assessing the value of acquiring new information. With this review we hope to bring an applied discipline (AM) to the attention of population ecologists and other researchers interested in the development of population models, while highlighting the relevance of population modeling in the AM framework for researchers and managers that are new to this area and are considering applying an AM approach.

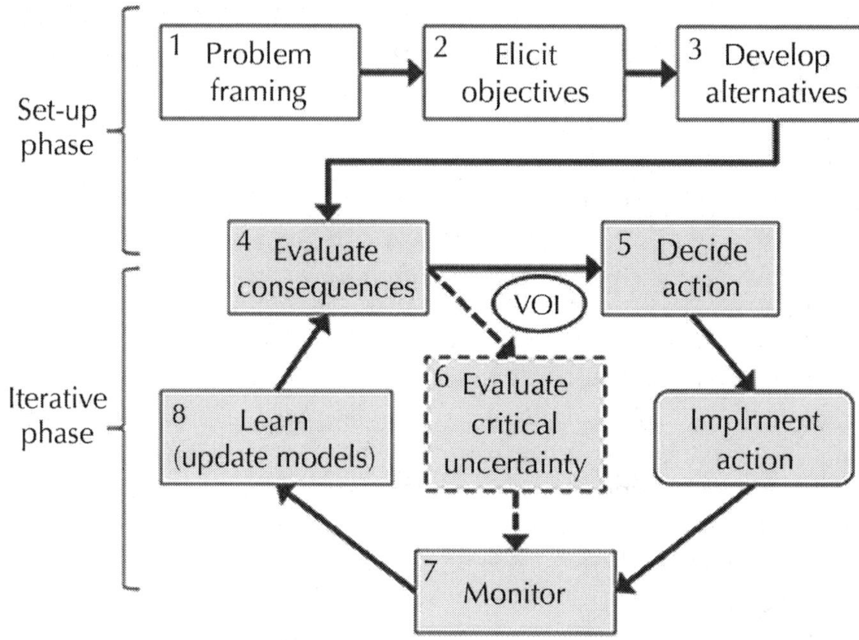

Figure 1: Diagram of the main steps in adaptive management, as an iterative form of structured decision making (adapted from Runge, 2011), highlighting the two main phases. VOI stands for "value of information analysis."

A QUICK TOUR OF ADAPTIVE MANAGEMENT: WHY AND HOW

AM is sometimes described as "learning while managing" but this broad expression may be misleading and a more careful definition is warranted: the key to an AM program is that there must be careful planning for learning from the first steps of the program definition (Walters, 1986; Parma et al., 1998; Williams et al., 2007, p. 31), rather than learning opportunistically from management mistakes as they occur, in a trial-and-error approach. Therefore, monitoring within AM must be specifically targeted to increase our knowledge of the system in a way that improves future management actions, in an iterative process of decision-action-monitoring-evaluation (Figure 1).

We do not delve in the details of AM itself in this short paper; the reader is directed to different sources in the literature: guidelines for managers (Williams et al., 2007), more methodological sources about quantitative methods (Walters and Hilborn, 1978; Williams, 2011c), overviews and reviews of different aspects of AM (Runge, 2011; Williams, 2011a; Westgate et al., 2013), including discussions about bridging the research-practice gap (Gregory et al., 2006; Allen and Gunderson, 2011) and circumstances when an AM approach is not warranted (Walters, 1986). The human or managerial dimension of the problem (i.e., long-term institutional commitment or proper stakeholder engagement; Williams et al., 2007, p. 18) can also influence the suitability of an AM approach; while recognizing its relevance, we do not discuss it here.

We note that the emphasis in AM is in facilitating a decision-making process which, depending on the case, may not require indulging in aspects of the ecology or population dynamics of a species if these are not necessary in order to make the decision. For example, some decisions simply require distinguishing between probabilities of success of different management options (e.g., McCarthy and Possingham, 2007) independently of the mechanistic explanation behind their performance; in such cases a population model, although informative and more portable to other situations, may not be necessary for AM to proceed.

Adaptive management can be seen as an iterative form of structured decision making (Runge, 2011; Gregory et al., 2012), a common framework for the application of decision theory grounded in value-focused thinking, which emphasizes the early identification of management objectives (Figure 1). The following two sections describe the role that population models can play in the different steps of this process, which consists of a set-up phase and an iterative phase.

SET-UP PHASE

Properly setting the context for the decision is the focus of the set-up phase, in which the management problem is discussed and consensus between stakeholders is reached regarding objectives and acceptable alternative actions. The consequences of actions are evaluated and decision trade-offs clarified so that a preferred action can be identified.

At the problem-framing stage [step 1, Figure 1], population models may be used to help understand the biological context of the problem, e.g., to evaluate the current status of a species, threatening factors, or the impact of harvest levels. Such models may also influence how objectives (and measures to gage them) are articulated [step 2]. A set of alternative management actions [step 3] is often elicited based on conceptual models of why there is a problem to be solved (e.g., different actions may address different hypotheses about reproductive failure; Runge et al., 2011); this step can sometimes be supported by population models (e.g., including/excluding the effect of a competitor on survival of the target species). Evaluating the consequences of alternative actions [step 4] may be based on expert opinion but would ideally use predictions from those models. The final step is to identify a preferred management action [step 5], which often involves assessing trade-offs (e.g., an action that provides higher expected population growth may also involve higher risk of a poor outcome) and may require using different decision tools depending on the complexity of the decision context (e.g., multi-criteria decision analysis, dynamic programming).

Alternative hypotheses and management actions can sometimes be based on different beliefs about population dynamics even without defining population models explicitly (e.g., disease persistence and resulting population growth rates; McDonald-Madden et al., 2010); parameterization may be based on expert elicitation during the set-up phase (Runge et al., 2011). In general, the structure chosen for the population model must take into account how it will be used to make a decision and to test alternative actions (Converse et al., 2013). It must therefore be able to reflect the set of hypotheses regarding the effect that alternative actions have on demography or population growth of the managed population, and provide good metrics for the objectives.

Objectives, model structures and uncertainties depend on the type of managed resource. For harvested populations, objectives are usually linked to maximizing long-term yield, by establishing optimal harvest rates. Early fisheries AM programs were based on a growth model with uncertain parameter values (e.g., intrinsic growth rate, Walters and Hilborn, 1976; density-dependence, Smith and Walters, 1981) or uncertain model structure (e.g., Ricker vs. Beverton-Holt stock-recruitment models, Walters and Hilborn, 1976). For waterfowl harvests in the US (Nichols et al., 1995; Johnson et al., 1997), alternative demographic models have been used to reflect relevant hypotheses about mortality being additive or compensatory (Williams et al., 1996) or the strength of density-dependence in recruitment (Johnson et al., 1997). In species conservation problems, objectives are usually related to population viability (Converse et al., 2013). This objective may be explicit (either maximizing viability or minimizing probability of extinction; Bakker and Doak, 2009) or defined indirectly through maximizing population abundance (Bearlin et al., 2002; Runge, 2013; Smith et al., 2013), number of populations (Moore et al., 2011) or population growth rate (McDonald-Madden et al., 2010), and may include thresholds (e.g., productivity above some level, Martin et al., 2010) or multiple competing objectives (e.g., harvest and conservation, Varley and Boyce, 2006; Smith et al., 2013). All these metrics require population models for their calculation. Again, the uncertainty that complicates decision-making may be related to demographic parameters (e.g., survival, Rout et al., 2009; Runge, 2013; or productivity, Smith et al., 2013) or model structure (Martin et al., 2010; Moore et al., 2011). The issue of population control is in a way inverse to that of conservation: managers attempt to keep the population of the problem species under some level, or even aim at eradicating it in the case of invasive species. AM has been proposed for the control of native species (e.g., overgrazing, Kaji et al., 2010; or predators, Martin et al., 2010) as well as alien invasive species (Parkes et al., 2006) and forestry/agricultural pests (Shea et al., 2002). Applications of AM to pest control in agriculture and forestry are still surprisingly rare given that these systems lend themselves well to experimental manipulation (Parma et al., 1998; Shea et al., 2002).

The population models defined during the setup phase are sometimes used in simulations to evaluate different management options before proceeding to the implementation of an AM program, by simulating the complete cycle of decision-action-monitoring-evaluation (Bearlin et al., 2002; Smith et al., 2013), an approach sometimes termed Management Strategy Evaluation (Bunnefeld et al., 2011). Demographic models can also be used in detailed simulations of population trajectories based on demographic parameters to evaluate management actions in terms of the resulting extinction risk, a process commonly known as "population viability analysis" or PVA (e.g., studies cited in Converse et al., 2013). A stronger integration of PVA into AM has been advocated as a way to explicitly link monitoring and management models in a single simulation framework ("Population Viability Management," Bakker and Doak, 2009).

ITERATIVE PHASE

The steps outlined in the previous section lead to a decision in a structured way. A key element in AM (particularly the "decision-theoretic" school of thought; Runge, 2011) is a subsequent iterative phase of monitoring to help resolve our uncertainty about decision-making, updating the belief in alternative hypotheses about the system, and adapting future actions accordingly (Figure 1).

This iterative process requires the articulation of critical uncertainty in the form of competing predictive population models [step 6]. This step strongly influences the design of the necessary monitoring program [step 7] so that it is targeted at resolving the critical uncertainties about the system that impede clear management decisions. Learning [step 8] comes from comparing the responses predicted by the competing population models to the response observed by monitoring the population; this allows updating the belief in the different hypotheses represented by these models. A key feature in AM is that we now adapt our subsequent decision-making in the light of the new knowledge, and enter a new iteration of the cycle.

Different actions are expected to generate different responses, and this often translates to different kinds and speeds of learning. Thus, there is a trade-off between the potential future benefits of learning from novel actions and data, against the immediate benefits of managing with what we know. A classic example from fisheries is a heavily fished population. We might assume a logistic growth model where we have good knowledge of the intrinsic growth rate but high uncertainty about carrying capacity. Maximum sustainable yield is a function of carrying capacity so managers may consider continuing with the intensive fishing or alternatively allowing the population to grow and approach carrying capacity by reducing fishing rates for a few years in order to reveal the population's true maximum sustainable yield (Walters and Hilborn, 1976). Two main approaches exist to deal with this type of trade-offs (Walters, 1986; Williams, 2011b): one can experiment actively (deviate from what appears the best option at a given time, in order to improve understanding so that future decisions can be improved—active adaptive management); or learn in a more passive fashion (act according to best current knowledge, but plan to learn from the outcome in order to improve future decisions—passive adaptive management). By contrast, we can also develop strategies that maximize distinction among hypotheses at the conclusion of the management program without any regard for management objectives (Probert et al., 2011).

Less structured approaches can also be used for decision-making, with AM proceeding on a suboptimal trajectory that still delivers learning alongside management (Williams et al., 2007, p. 33; Converse et al., 2013). Whenever possible, optimizing the decision-making process will improve the predicted benefits of an AM approach. For both active and passive AM, quantitative methods exist to provide an optimal schedule of actions so that an overall objective is maximized for a given timeframe. Optimization methods most commonly assume that populations possess first-order Markovian dynamics, such that all feedback from the current action is observed in the next time step. Any delayed responses need to be incorporated as state variables, which increase the complexity of the problem.

Stochastic dynamic programming has been used widely to calculate optimal schedules (e.g., Walters and Hilborn, 1976; Johnson et al., 1997; McCarthy and Possingham, 2007), and uses discrete transition models of system and knowledge change over time. This technique is typically constrained to small state-spaces and simple knowledge constructions, involving discrete approximation of continuous population models. Walters (1986) recommended Kalman filters to approximate more complex (but still discrete) knowledge transitions, and wide-sense dual control algorithms to approximate optimal management schedules. Other methods can accommodate continuity in knowledge states for a small number of distinct alternative population models (Williams, 2011c). For more complex structures, artificial intelligence methods and algorithms for discretized belief Markov Decision Processes (Fackler and Pacifici, 2014), Partially Observable Markov Decision Processes (Nicol et al., 2013) and Mixed Observability Markov Decision Processes (Chadès et al., 2012) show promise for developing near-optimal management schedules. These approaches assume discrete population model transitions between time steps; thus appropriate discretization of population models will require a trade-off of predictive accuracy against the computational burden of optimizing over a large state-space, although in some cases few states really matter (Nicol and Chadès, 2012).

VALUE OF INFORMATION ANALYSIS

Within the idea of "learning while managing" sits the concept of acquiring new information to improve the decision-making process. A set of quantitative tools from decision theory, collectively called "Value of Information analysis" (VOI), allow investigation of the value of collecting additional information to resolve or reduce uncertainty (Runge et al., 2011). The most common techniques are Expected Value of Perfect Information (EVPI; resolving all uncertainty), Expected Value of Partial Information (EVPI; resolving a subset of uncertainty, e.g., one particular hypothesis) and Expected Value of Sample Information (EVSI; for a data sample of a given size).

Analyzing the value of information can help: (1) concentrate the learning to specific aspects of uncertainty that will improve decision-making more quickly (high EVPI) and/or avoid doing research on those aspects that do not contribute significantly (low EVPI); (2) decide the amount of information that has to be collected to achieve a given level of improvement in decision-making (EVSI); or (3) indicate when overall it is not worth attempting to learn about the system (low EVPI) or at least given the available resources or capabilities to monitor (low EVSI for realistic sample sizes). After all, the value of AM hinges on having high value of information and the means to acquire enough data to achieve sufficient power (Runge, 2011). VOI analysis could be conducted before committing to an AM program and at the beginning of each management cycle if our knowledge has changed substantially; sequential VOI analysis has been proposed (Griffin et al., 2010) but we are not aware of any application in environmental management.

Population models can play a key role in conducting VOI analyses, which require quantifying the effect of alternative management actions on some parameter(s) of interest, under alternative hypotheses about the system. Population modelers must therefore translate these hypotheses into different model structures or parameter values so that the impact of actions can be predicted. For example, in Runge et al. (2011) expert elicitation provided expected survival probabilities under several reproductive failure hypotheses and management strategies; if enough data were available these probabilities could be estimated using alternative models. Furthermore, VOI analysis is based on demographic parameters or population growth, but the managers that are in a position to decide whether extra data collection is worth its cost often prefer to work in terms of the population impact of these demographic processes, e.g., persistence or extinction risk. Population models can also be used to convert between these currencies, e.g., translating a change in expected survival probability due to resolving some uncertainty to an expected change in extinction risk. Population viability analysis (PVA), which includes a demographic model at their core, is particularly useful for this purpose. In summary, we believe VOI analysis can play an important role in informing research agendas for population modeling since it may indicate which aspects of the demography of a population are worth investigating.

CONCLUSIONS

Adaptive management represents ecological intervention with a structured plan for learning about key uncertainties in a natural system of interest (Parma et al., 1998). Issues of harvest, pest management and conservation all involve fundamental objectives concerned with future population status: to maintain, reduce or increase their levels, respectively. Since demographic parameters govern the fate of the population under management, population models can provide the necessary predictive link between management alternatives and objectives (Converse et al., 2013) and can therefore be central to adaptive management programs. Calls to extend the use of AM in decisions regarding wildlife populations are common in the literature (Parma et al., 1998; Shea et al., 2002; Williams et al., 2007; Runge, 2011; Williams, 2011a; Converse et al., 2013). We believe a more widespread application of adaptive management in natural resource management and biodiversity conservation will provide opportunities for stronger collaboration between population modelers, managers and other stakeholders, to achieve a better integration of science and management.

ACKNOWLEDGMENTS

This work was supported by the ARC Centre of Excellence for Environmental Decisions and the National Environment Research Program (NERP) Decisions Hub. We would like to thank Iadine Chadès for useful discussion regarding optimization methods.

REFERENCES

1. Allen, C. R., and Gunderson, L. H. (2011). Pathology and failure in the design and implementation of adaptive management. J. Environ. Manage. 92, 1379–1384. doi: 10.1016/j.jenvman.2010.10.063
2. Bakker, V. J., and Doak, D. F. (2009). Population viability management: ecological standards to guide adaptive management for rare species. Front. Ecol. Environ. 7, 158–165. doi: 10.1890/070220
3. Bearlin, A. R., Schreiber, E. S. G., Nicol, S. J., Starfield, A. M., and Todd, C. R. (2002). Identifying the weakest link: simulating adaptive management of the reintroduction of a threatened fish. Can. J. Fish. Aquat. Sci. 59, 1709–1716. doi: 10.1139/f02-140
4. Bunnefeld, N., Hoshino, E., and Milner-Gulland, E. J. (2011). Management strategy evaluation: a powerful tool for conservation? Trends Ecol. Evol. 26, 441–447. doi: 10.1016/j.tree.2011.05.003
5. Caswell, H. (2001). Matrix Population Models: Construction, Analysis and Interpretation. Sunderland: Sinauer Associates.
6. Chadès, I., Carwardine, J., Martin, T. G., Nicol, S., Sabbadin, R., and Buffet, O. (2012). "MOMDPs: a solution for modelling adaptive management problems," in Twenty-Sixth AAAI Conference on Artificial Intelligence (Toronto, ON: AAAI Press).
7. Converse, S. J., Moore, C. T., and Armstrong, D. P. (2013). Demographics of reintroduced populations: estimation, modeling, and decision analysis. J. Wildl. Manage. 77, 1081–1093. doi: 10.1002/jwmg.590
8. Fackler, P., and Pacifici, K. (2014). Addressing structural and observational uncertainty in resource management. J. Environ. Manage. 133, 27–36. doi: 10.1016/j.jenvman.2013.11.004

9. Gerber, L. R., Beger, M., McCarthy, M. A., and Possingham, H. P. (2005). A theory for optimal monitoring of marine reserves. Ecol. Lett. 8, 829–837. doi: 10.1111/j.1461-0248.2005.00784.x
10. Gregory, R., Failing, L., Harstone, M., Long, G., Mcdaniels, T., and Ohlson, D. (2012). Structured Decision Making: A Practical Guide to Environmental Management Choices. Chichester, UK: John Wiley & Sons. doi: 10.1002/9781444398557
11. Gregory, R., Ohlson, D., and Arvai, J. (2006). Deconstructing adaptive management: criteria for applications to environmental management. Ecol. Appl. 16, 2411–2425. doi: 10.1890/1051-0761(2006)016[2411:DAMCFA]2.0.CO;2
12. Griffin, S., Welton, N. J., and Claxton, K. (2010). Exploring the research decision space: the expected value of information for sequential research designs. Med. Decis. Making 30, 155–162. doi: 10.1177/0272989X09344746
13. Johnson, F. A., Moore, C. T., Kendall, W. L., Dubovsky, J. A., Caithamer, D. F., Kelley, J. R., et al. (1997). Uncertainty and the management of mallard harvests. J. Wildl. Manage. 61, 202–216. doi: 10.2307/3802429
14. Kaji, K., Saitoh, T., Uno, H., Matsuda, H., and Yamamura, K. (2010). Adaptive management of sika deer populations in Hokkaido, Japan: theory and practice. Popul. Ecol. 52, 373–387. doi: 10.1007/s10144-010-0219-4
15. Martin, J., O'connell, A. F., Kendall, W. L., Runge, M. C., Simons, T. R., Waldstein, A. H., et al. (2010). Optimal control of native predators. Biol. Conserv. 143, 1751–1758. doi: 10.1016/j.biocon.2010.04.023
16. McCarthy, M. A., and Possingham, H. P. (2007). Active adaptive management for conservation. Conserv. Biol. 21, 956–963. doi: 10.1111/j.1523-1739.2007.00677.x
17. McDonald-Madden, E., Probert, W. J. M., Hauser, C. E., Runge, M. C., Possingham, H. P., Jones, M. E., et al. (2010). Active adaptive conservation of threatened species in the face of uncertainty. Ecol. Appli. 20, 1476–1489. doi: 10.1890/09-0647.1
18. Moore, C. T., and Conroy, M. J. (2006). Optimal regeneration planning for old-growth forest: addressing scientific uncertainty in endangered species recovery through adaptive management. Forest Sci. 52, 155–172.

19. Moore, C. T., Fonnesbeck, C. J., Shea, K., Lah, K. J., McKenzie, P. M., Ball, L. C., et al. (2011). An adaptive decision framework for the conservation of a threatened plant. J. Fish Wildl. Manage. 2, 247–261. doi: 10.3996/012011-JFWM-007
20. Nichols, J. D., Johnson, F. A., and Williams, B. K. (1995). Managing North-American waterfowl in the face of uncertainty. Annu. Rev. Ecol. Syst. 26, 177–199. doi: 10.1146/annurev.es.26.110195.001141
21. Nicol, S., Buffet, O., Iwamura, T., and Chadès, I. (2013). "Adaptive management of migratory birds under sea level rise," in Twenty-Third International Joint Conference on Artificial Intelligence (Beijing: AAAI Press).
22. Nicol, S., and Chadès, I. (2012). Which states matter? an application of an intelligent discretization method to solve a continuous POMDP in conservation biology. PLoS ONE 7:e28993. doi: 10.1371/journal.pone.0028993
23. Parkes, J. P., Robley, A., Forsyth, D. M., and Choquenot, D. (2006). Adaptive management experiments in vertebrate pest control in New Zealand and Australia. Wildl. Soc. Bull. 34, 229–236. doi: 10.2193/0091-7648(2006)34[229:AMEIVP]2.0.CO;2
24. Parma, A. M., Amarasekare, P., Mangel, M., Moore, J., Murdoch, W. W., Noonburg, E., et al. (1998). What can adaptive management do for our fish, forests, food, and biodiversity? Integr. Zool. 1, 16–26.
25. Probert, W. J. M., Hauser, C. E., McDonald-Madden, E., Runge, M. C., Baxter, P. W. J., and Possingham, H. P. (2011). Managing and learning with multiple models: objectives and optimization algorithms. Biol. Conserv. 144, 1237–1245. doi: 10.1016/j.biocon.2010.07.031
26. Regan, H. M., Colyvan, M., and Burgman, M. A. (2002). A taxonomy and treatment of uncertainty for ecology and conservation biology. Ecol. Appl. 12, 618–628. doi: 10.1890/1051-0761(2002)012[0618:ATATOU]2.0.CO;2
27. Ricker, W. E. (1954). Stock and recruitment. J. Fish. Res. Board Can. 11, 559–623. doi: 10.1139/f54-039
28. Rout, T. M., Hauser, C. E., and Possingham, H. P. (2009). Optimal adaptive management for the translocation of a threatened species. Ecol. Appl. 19, 515–526. doi: 10.1890/07-1989.1

29. Runge, M. C. (2011). An introduction to adaptive management for threatened and endangered species. J. Fish Wildl. Manage. 2, 220–233. doi: 10.3996/082011-JFWM-045
30. Runge, M. C. (2013). Active adaptive management for reintroduction of an animal population. J. Wildl. Manage. 77, 1135–1144. doi: 10.1002/jwmg.571
31. Runge, M. C., Converse, S. J., and Lyons, J. E. (2011). Which uncertainty? using expert elicitation and expected value of information to design an adaptive program. Biol. Conserv. 144, 1214–1223. doi: 10.1016/j.biocon.2010.12.020
32. Sainsbury, K. J. (1991). Application of an experimental approach to management of a tropical multispecies fishery with highly uncertain dynamics. ICES Marine Sci. Symp. 193, 301–320.
33. Shea, K., Possingham, H. P., Murdoch, W. W., and Roush, R. (2002). Active adaptive management in insect pest and weed control: intervention with a plan for learning. Ecol. Appli. 12, 927–936. doi: 10.1890/1051-0761(2002)012[0927:AAMIIP]2.0.CO;2
34. Smith, A. D. M., and Walters, C. J. (1981). Adaptive management of stock-recruitment systems. Can. J. Fish. Aquat. Sci. 38, 690–703. doi: 10.1139/f81-092
35. Smith, D. R., McGowan, C. P., Daily, J. P., Nichols, J. D., Sweka, J. A., and Lyons, J. E. (2013). Evaluating a multispecies adaptive management framework: must uncertainty impede effective decision-making? J. Appl. Ecol. 50, 1431–1440. doi: 10.1111/1365-2664.12145
36. Varley, N., and Boyce, M. S. (2006). Adaptive management for reintroductions: updating a wolf recovery model for Yellowstone National Park. Ecol. Modell. 193, 315–339. doi: 10.1016/j.ecolmodel.2005.09.001
37. Walters, C. (1986). Adaptive Management of Renewable Resources. New York, NY: Macmillan Publishing Co.
38. Walters, C. J., and Hilborn, R. (1976). Adaptive-control of fishing sysetems. J. Fish. Res. Board Can. 33, 145–159. doi: 10.1139/f76-017
39. Walters, C. J., and Hilborn, R. (1978). Ecological optimization and adaptive management. Annu. Rev. Ecol. Syst. 9, 157–188. doi: 10.1146/annurev.es.09.110178.001105

40. Westgate, M. J., Likens, G. E., and Lindenmayer, D. B. (2013). Adaptive management of biological systems: a review. Biol. Conserv. 158, 128–139. doi: 10.1016/j.biocon.2012.08.016
41. Williams, B. K. (2011a). Adaptive management of natural resources—framework and issues. J. Environ. Manage. 92, 1346–1353. doi: 10.1016/j.jenvman.2010.10.041
42. Williams, B. K. (2011b). Passive and active adaptive management: approaches and an example. J. Environ. Manage. 92, 1371–1378. doi: 10.1016/j.jenvman.2010.10.039
43. Williams, B. K. (2011c). Resolving structural uncertainty in natural resources management using POMDP approaches. Ecol. Modell. 222, 1092–1102. doi: 10.1016/j.ecolmodel.2010.12.015
44. Williams, B. K., Johnson, F. A., and Wilkins, K. (1996). Uncertainty and the adaptive management of waterfowl harvests. J. Wildl. Manage. 60, 223–232. doi: 10.2307/3802220
45. Williams, B. K., Szaro, R. C., and Shapiro, C. D. (2007). Adaptive Management: The U.S. Department of the Interior Technical Guide. Adaptive Management Working Group. Washington, DC: Department of the Interior. Available online at: http://www.doi.gov/ppa/upload/TechGuide-WebOptimized-2.pdf

Chapter 8

Fire Induced Reproductive Mechanisms of a Phoricarpos (Caprifoliaceae) Shrub after Dormant Season Burning

John Derek Scasta[1], David M Engle[2], Ryan N Harr[3], and Diane M Debinski[4]

[1]Department of Ecosystem Science and Management, University of Wyoming, Laramie 82071, WY, USA

[2]Department of Natural Resource Ecology and Management, Oklahoma State University, Stillwater, OK, USA

[3]Iowa Department of Natural Resources, Des Moines, IA, USA

[4]Department of Ecology, Evolution, and Organismal Biology, Iowa State University, Ames, IA, USA

ABSTRACT

Background

Symphoricarpos, a genus of the Caprifoliaceae family, consists of about 15 species of clonal deciduous shrubs in North America and 1 species endemic to China. In North American tallgrass prairie, *Symphoricarpos orbiculatus* (buckbrush) is the dominant shrub often forming large colonies via sexual and asexual reproductive mechanisms. *Symphoricarpos* shrubs, in particular *S. orbiculatus,* use a unique sexual reproductive mechanism known as layering where vertical stems droop and the tips root upon contact with the soil. Because of conflicting societal values of *S. orbiculatus* for conservation and agriculture and the current attempt to restore historical fire regimes, there is a need for basic research on the biological response of *S. orbiculatus* to anthropogenic burning regimes.

Results

From 2007 through 2013 we applied prescribed fires in the late dormant season on grazed pastures in the Grand River Grasslands of Iowa. From 2011 to 2013, we measured how *S. orbiculatus* basal resprouting and layering stems were affected by patchy fires on grazed pastures, complete pasture fires on grazed pastures or fire exclusion without grazing for more than three years. We measured ramet height, ramet canopy diameter, stems per ramet, ramets per 100 m^2, and probability of new layering stems 120 days after fire. Height in burned plots was lower than unburned plots but *S. orbiculatus* reached ~ 84% of pre-burn height 120 days after fire. Stems per ramet were 2x greater in the most recently burned plots due to basal re-sprouting. Canopy diameter and density of ramets was not affected by time since fire, but burned pastures had marginally lower densities than plots excluded from fire ($P = 0.07$). Fire triggered new layering stems and no new layering stems were found in plots excluded from fire.

Conclusions

The mechanisms of both basal sprouting and aerial layering after fire suggest *S. orbiculatus* is tolerant to dormant season fires. Furthermore, dormant season fires, regardless if they were patchy fires or complete pasture fires, did not result in mortality of *S. orbiculatus*. Dormant season fires can reduce *S. orbiculatus* structural dominance and maintain lower ramet densities but also trigger basal resprouting and layering.

BACKGROUND

Global vegetation distribution patterns are largely driven by climatic constraints and disturbance regimes (Clark [1991]; Collins [2000]). Grasslands on several continents were regulated by the developmental disturbances of fire, large ungulate grazing, and the synergistic interaction of the two (Fuhlendorf and Engle [2001]). The vegetation composition of these fire prone grasslands includes shrubby angiosperm species that persist after fire due to vegetative regeneration mechanisms rather than recruitment (Bond and Midgley [2003]). In North America, conversion to cropland and suppression of fire has rendered tallgrass prairies as one of the most threatened ecosystems globally (Sampson and Knopf [1994]). Prior to European settlement, tallgrass prairie was prone to burning every three to five years (Collins [2000]). Consequently, the exclusion of regular fire in the remaining tallgrass prairie alters plant succession toward a shrub dominated community and shrub encroachment is a global issue (Collins and Adams [1983]; Collins [1992]; Shortt and Vamosi [2012]).

Shrubs in these fire prone ecosystems utilize a resprouting life strategy that results in tradeoffs between growth and reproduction because resprouting shrubs often have smaller seeds, poor reproduction from seeds, and short-lived seed banks (Bond and Wilgen [1996]; Kruger et al.[1997]). The cost of energy allocation belowground to survive fire also reduces growth rate (Bond and van Wilgen [1996]). Subsequently, form follows function, and resprouting species typical of disturbance prone ecosystems are multi-stemmed and short statured to optimize rapid recovery. In contrast, recruiting species are typically single-stemmed and taller, an advantage for the competition for light and dispersion of seed (Bellingham and Sparrow [2000]; Bond and Midgley[2003]). Researchers have focused more on seedling ecology than resprouting mechanisms as a plant life history trait, although neither regeneration strategy is mutually exclusive (Olson and Platt[1995]; Higgins et al. [2000]; Bond and Midgley [2003]). Furthermore, basal resprouting mechanisms are not uniform within a functional group type (i.e., not all shrubs resprout basally or epicormically) or even within genera and are less understood than seedling recruitment (Enright and Goldblum [1999]; Bellingham and Sparrow [2000]; Ansley and Rasmussen [2005]; Winter et al. [2011]).

Symphoricarpos is a genus of the honeysuckle or Caprifoliaceae family, consisting of about 15 species of clonal deciduous shrubs in North America and 1 species endemic to China (*S. sinensis*) (Theis et al. [2008]). In North America, *Symphoricarpos* is one of the most broadly distributed shrub genera, and *S. orbiculatus* has been described as the dominant shrub of the tallgrass prairie (Holechek et al. [2001]; USDA Plant Database [2013]; Scasta [2014]). Asexual spread is from rhizomes and rooting of aerial stems (henceforth layering) and sexual spread is from seeds (Pelton [1953]; Hullick and Manske [2006]; Nesmith et al. [2006]). Layering occurs when vertical stems droop horizontally and produce adventitious roots when they come into contact with the soil. While the layering mechanism has been studied from a physical disturbance standpoint, it has been largely neglected in the fire ecology literature (Hartmann and Kester [1975]; Deb and Pogener [2012]). Furthermore, *S. orbiculatus* has been characterized as a major invader of unburned prairie but the role of fire exclusion and the application of fire with livestock grazing are not well understood for this species (Bragg and Hulbert [1974]; Stubbendieck et al. [2003]). However, because *S. orbiculatus* is important for a wide range of wildlife species, there are conflicting views on the necessity and desirability of reducing *S. orbiculatus* (Korschgen [1962]; Guthrey [1980]; Brennan [1991]; Soper et al. [1993]; Harrell et al. [2001]; Stubbendieck et al.[2003]).

Quantifying *S. orbiculatus* reproductive mechanisms after fire will yield basic biological insights to assist managers in appropriately applying disturbance regimes and potentially develop alternative strategies to mitigate encroachment. Our objectives were to assess how *S. orbiculatus* vegetative regeneration and reproduction is affected by fire exclusion and late dormant-season fire applied every third year across entire grazed pastures or applied annually on a third of a grazed pasture. Prescribed burning in the late dormant season is the prescribed fire season most common in North American grasslands and is limited to after snow melt and before vegetation greenup. We hypothesized that *S. orbiculatus* is tolerant to dormant season burning and that the reproductive mechanisms of basal sprouting and aerial layering would be dependent on time since fire and serve as the key mechanisms of tolerance. To test our hypothesis we quantified vegetative response to prescribed fire by measuring ramet height, ramet canopy diameter, stems per ramet, ramets per 100 m^2, and probability of new layering stems.

METHODS

Study Location and Design

The study was conducted from 2011–2013 in tallgrass prairie in Ringgold County, Iowa in the Grand River Grasslands (GRG) of Iowa, USA (40°34′N, 94°10′W). The GRG is in the glaciated plains of the Central Tallgrass Prairie Ecoregion and has been described as one of the premier places to restore a functioning tallgrass prairie ecosystem (Missouri Department of Conservation[2010]). Annual precipitation was 968 mm in 2011 (+97 mm above long-term mean), 798 mm in 2012 (− 73 mm below long-term mean), 874 mm in 2013 (+3 mm above long-term mean and averaged 870 mm during the three year study (+1 mm above long-term mean) (Iowa Environmental Mesonet [2014]). Soils are loess hills with glacial till side slopes with slopes in some areas exceeding 9%. Subsoils have high clay content ranging from 42 to 48% and native vegetation was tallgrass prairie (USDA NRCS [2013]). Herbaceous vegetation across all study pastures was dominated by perennial C4 graminoids (*Andropogon gerardii* (big bluestem),*Schizacyrium scoparium* (little bluestem), and *Sorghastrum nutans* (Indiangrass) with a component of exotic C3 graminoids and legumes (*Schedonorus arundinacea* (tall fescue), *Bromus inermis*(smooth brome), *Lotus corniculatus* (birdsfoot trefoil), and *Trifolium spp.* (clover). In these study pastures, *S. orbiculatus* was the most common shrub regardless of the spatial scale of assessment and occurred more often than any other shrub species including *Rhus glabra* (smooth sumac),*Rosa multiflora* (multiflora rose), *Cornus drummondii* (dogwood), *Prunus spp.* (plum) and *Rubus spp.* (raspberry) (Scasta [2014]).

Fire Effects Sampling

Three treatments were applied and assessed: 1) Patch-burn grazing (PBG) or burning one-third of a pasture annually (the patch) with cattle having full access to the pasture, 2) Graze and burn (GAB) or burning pastures completely every three years with cattle having full access to the pasture, and 3) Unburned or no fire for over three years and no cattle grazing during that period. Grazing was seasonal from late-April to October with mature Angus cows. Mean (± SE) size of each pasture was 27 ± 3 ha and mean (± SE) stocking rate was 2.3 ± 0.2 AUM ha^{-1}. All prescribed fires were conducted within a four week window in the late dormant season (March 8 to April 4) prior to the emergence of *S. orbiculatus* leaves. Mean fire weather data for the available fires was 47% relative humidity, 9 km per hr, and 18°C. The fire and grazing treatments were in place prior to the initiation of sampling in 2011, so all patches and pastures had a consistent elapsed time since fire (0.3, 1, 2 years since fire) (Table 1). The designation of 0.3 years since fire indicates that samples were measured four months after fire. We had less control over the unburned treatments and elapsed time since fire for those pastures was more than three years but the exact number of years is unknown. When we returned in 2013 to conduct a second assessment of unburned pastures, they had all either been burned or mowed so no follow up measurements were possible.

Table 1: Schedule of prescribed fire and measurements of*Symphoricarpos orbiculatus*in the Grand River Grasslands, USA, 2007-2013

	Year						
Treatment	2007	2008	2009	2010	2011	2012	2013
Patch-burn grazed pastures (3 pastures total delineated into 3 patches per pasture)							
Pyland North							
Patch A	fire	-	-	fire	-, m1, m2	-, m1	fire, m1, m2

Patch B	-	fire	-	-	fire, m1, m2	-, m1	-, m1, m2
Patch C	-	-	fire	-	-, m1, m2	fire, m1	-, m1, m2
Pyland South							
Patch A	fire	-	fire*	fire	-, m1, m2	-, m1	fire, m1, m2
Patch B	-	fire	fire*	-	fire, m1, m2	-, m1	-, m1, m2
Patch C	-	-	fire	-	-, m1, m2	fire, m1	-, m1, m2
Ringgold South							
Patch A	fire	-	-	fire	-, m1, m2	-, m1	fire, m1, m2
Patch B	-	fire	-	-	fire, m1, m2	-, m1	-, m1, m2
Patch C	-	-	fire	-	-, m1, m2	fire, m1	-, m1, m2
Graze-and-burn pastures (3 pastures total)							
Gilleland	-	-	fire	-	-, m1	fire, m1	-, m1
Lee Trail	-	-	fire	-	-, m1	fire, m1	-, m1
Pyland West	-	-	fire	-	-, m1	fire, m1	-, m1
Unburned pastures (3 pastures total)							
Hog Farm	-	-	-	-	-, m1, m2	-	-, d
Kellerton South	-	-	-	-	-, m1, m2	-	-, d
Richardson West	-	-	-	-	-, m1, m2	-	-, d

Fire: –=no fire; fire=prescribed fire; fire*=escaped out of the intended burn patch and burned remainder of pasture; d=disrupted (we had less control over the unburned treatment and when we returned in 2013 to measure a second assessment of unburned pastures, they had all either been burned or mowed). (d=disrupted).

Measurements: m_1 =Symphoricarpos orbiculatus maximum ramet height, ramet maximum canopy diameter, stems per ramet measured as number of vertical live stems arising from the same root crown; m_2 = new aerial layering stems.

Treatments: Patch-burn grazing (PBG) where one-third of a pasture (the patch) is burned and cattle have full access to the pasture, Graze and burn (GAB) where the entire pasture is burned every third year (2012 was the burn year) and cattle have full access to the pasture, and Not burned for more than three years with no grazing. Treatments were in place starting in 2007 so all patches and pastures had a consistent elapsed time since fire (0.3, 1, 2 years since fire).

Scasta et al.

Scasta et al. Botanical Studies 2014 55:80 doi:10.1186/s40529-014-0080-4

Fire effects measurements were conducted from 2011 to 2013 in six plots per pasture controlling for catena and soil type (Debinski et al. [2011]). Plots were all permanent and ranged from 16 m^2(16 m × 1 m) to 350 m^2 (25 m × 14 m). Plots were variable in size due to the variable density of*S. orbiculatus* ramets. If *S. orbiculatus* was not present within a permanent plot, we assessed 25 m north or south of the permanent marking pins for *S. orbiculatus* presence. Three additional pastures that had not been burned in more than three years were included for fire effects measurements and had three plots per pasture (Table 1). Thus, to ensure adequate sample size and in the sake of time, we sampled plots for a minimum of ten ramets to a maximum of fifty ramets in each plot. Given the clonal nature of *S. orbiculatus* and the intermingling of lateral vegetative structures among adjacent clones, it was not possible to distinguish individual clones (aka, genets). Therefore, we conducted measurements at the ramet level. We define ramets as individual plants or clonal fragments in the colony that are rooted and may have originated from either seed or rooted nodes from aerial layering stems that may or may not be connected to other ramets (Nesmith et al. [2006]) (Figure 1). Ramet measurements were conducted approximately 120 days after fire. Maximum height of the ramet was measured as the tallest individual stem of a ramet. Maximum canopy diameter was measured as the longest horizontal axis of a ramet. Stems per ramet was measured as the total number of vertical live stems arising from the same root crown and is a reflection of the regeneration from basal resprouts. Ramets per 100 m^2 were measured by noting the total number of ramets in a plot and then converting that to 100 m^2, a reflection of alterations to ramet density from fire-induced mortality. New aerial layering stems were measured as vertical stems arising from the root crown but having a horizontal orientation but not yet rooting at the tip (Figure 1). All measurements for fire effects were conducted in 2011, 2012 and 2013 except presence/absence of new

Parameter	Estimate	Standard Error	Likelihood ratio 95% Confidence limits		Pvalue
Intercept	3.420	1.077	1.537	5.858	0.002
Time Since Fire	−1.330	0.403	−2.235	−0.622	0.001

Maximum likelihood parameter estimates for time since fire and probability of the presence of new aerial layering stems based on the logit function and binomial presence/absence data. Based on the exponentiation of the parameter estimate, the odds of having a layering stem present improve 3.77x for every year closer to the burn year.

Scasta et al.

Scasta et al. Botanical Studies 2014 55:80 doi:10.1186/s40529-014-0080-4

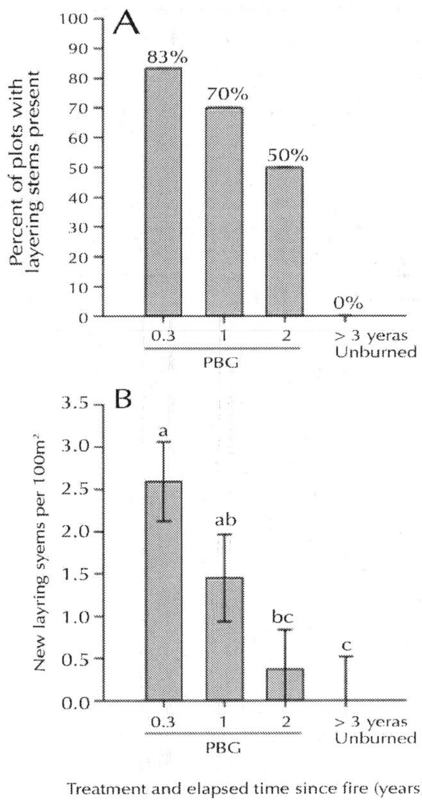

Figure 3: Effect of elapsed time since fire on new*Symphoricarpos orbiculatus*horizontal layering stems based on percent of plots with new layering

stems (A), and new layering stem density (B) in the unburned pastures and the Patch-burn grazed (PBG) pastures in in the Grand River Grasslands of Iowa, USA, 2011 and 2013.Letters indicate differences ($P \leq 0.05$) based on a mixed effects model.

DISCUSSION

Our study examining the regeneration and reproduction of *S. orbiculatus* after fire indicates that this shrub is tolerant of low-intensity prescribed fires conducted in the late dormant season or early spring prior to emergence of leaves. *S. orbiculatus* rapidly regenerated from basal resprouts, recovering to ~ 84% of its pre-fire height 120 days after fire with no apparent mortality. *S. orbiculatus* used both basal resprouting and aerial layering as reproduction strategies to recover after the disturbance of fire. Ramet height increased with time since fire, but stems per ramet and aerial layering probability decreased with time since fire. Initially, fire physically altered shrub structure as measured by ramet height and stems per ramet. However, the initial structural change was only temporary as the trend indicated that as time elapses past four years, physical shrub structure becomes similar to unburned pastures, similar to *Artemisia filifolia* (sand sagebrush) (Winter et al. [2011]).

The rapid structural regeneration of *S. orbiculatus* within the burn-year growing season is similar to other species that display rapid regrowth the first year with a declining rate of growth in the next three to four years (Gratani and Amadori [1991]). The rate of recovery for *S. orbiculatus* was slower than *Prosopis glandulosa* (mesquite) that recovered to pre-burn heights within one year after fire (Drewa [2003]). A similar study in tallgrass prairie suggested that *R. glabra* rapidly recovers to its pre-burn height and fire can result in greater plant heights than in unburned areas for both *P. glandulosa* and *R. glabra* (Drewa [2003]; Hajny et al. [2011]). However, fire did not appear to increase *S. orbiculatus* height in our study. In more arid environments with coarse sandy soils, *A. filifolia* and *Quercus havardii* (sand shinnery oak) recovered to only ~ 55% of pre-burn height within the first year of a burn (Harrell et al. [2001]; Winter et al. [2011]). However, *S. orbiculatus* is similar to *A. filifolia* and *Q. havardii* because it took three years or more for all three species to recover to pre-burn heights. Considering these variable responses, fire only altered the physical structure of these fire-tolerant shrubs for a relatively short period, typically less than four years, and these shrubs regenerated at different rates depending on climate and soil limitations (Olson and Platt [1995]; Iwasa and Kubo [1997]).

Because *S. orbiculatus* seeds have little to no potential to form a long lived soil seed bank (Hidayata et al. [2001]) and the rapid regeneration after fire reported herein, vegetative regeneration appears to be the primary recovery mechanism of *S. orbiculatus* to dormant season prescribed fires. In context, the tallgrass prairie is a disturbance driven ecosystem that historically had a fire return interval of three to five years (Collins [2000]). The life history tradeoff between resprouting and seedling recruitment is largely determined by disturbance regimes and tallgrass prairie was historically disturbed regularly by fire (Wells [1969]; Bellingham and Sparrow [2000]; Collins [2000]). Episodic resprouting events triggered by fire are followed by extended periods of self-thinning as indicated by the relationship with elapsing time since fire and declining stems per ramet in our study (Clark [1991]; Hodgkinson [1998]).

The low growing and multi-stemmed architecture of *S. orbiculatus* is also reflective of the disturbance regime of tallgrass prairie and its ability to rapidly recover after fire (Midgley [1996]; Bellingham and Sparrow [2000]). However, vegetative reproduction of *S. orbiculatus* after fire is not limited to basal resprouting regeneration but also appears to include aerial layering. The initiation of aerial layering has also been reported to be triggered by physical disturbances for *Rubus trivialis* (dewberries) from grazing, for *Oplopana horridus* (devil's club) from recent logging activity, and for *Cercocarpus* spp. (mountain mahogany) from grazing and possibly fire (Miller[1964]; Abrahamson [1975]; Lantz and Antos [2002]). The absence of cattle grazing activity as a source of physical disturbance in the unburned pastures could also help explain the lack of new *S. orbiculatus* layering stems (Abrahamson [1975]). In the Douglas-Fir forests of Oregon, USA, burning did not increase aerial layering but did increase basal resprouting of *Acer circinatum* (vine maple) while the physical disturbance of thinning and falling branches stimulated layering (O'Dea et al. [1995]). However, in this old growth forest ecosystem the historic fire return interval is estimated to be 230 years compared to our study area that burned every three to five years (Agee [1993]).

Our results quantifying the ability of *S. orbiculatus* to regenerate after low-intensity prescribed fires conducted in the late dormant season or early spring is similar to other *Symphoricarpos*species, although not every species is noted to be tolerant of fire. *S. occidentalis* (western snowberry) and *S. albus* (common snowberry) are highly tolerant to fire due to vegetative regeneration mechanisms (Mclean [1969]; Anderson and Bailey [1979]; Morgan and Neuenschwander [1988]; Romo et al. [1993]; Youngblood et al. [2006]; Hauser [2007]). In contrast, *S. oreophilus* (mountain snowberry), *S. longiflorus* (longflower snowberry), and *S. mollis*(creeping snowberry) have been reported to be only low to moderately resilient to fire (Bartos and Mueggler [1982]; Snyder [1991]; Aleksoff [1999]; McWilliams [2000]; McWilliams [2005]; Knapp et al. [2006]; Rocca [2009]). Interestingly, even if a species is not highly tolerant of fire, it still may be fire dependent as *S. oreophilus* occurred only on burned sites (Poreda and Wullstein[1994]). This variable response to fire within the genus is attributed to species distributions and biogeographic disturbance patterns because disturbance frequency is a major determinant of resprouting strength (Westman and O'Leary [1986]; USDA Plant Database [2013]).

In contrast to our study, other studies have reported reductions in *S. orbiculatus* with fire. Late spring burning in tallgrass prairie reduced *S. orbiculatus* due to lower carbohydrate reserves because plants had already leafed out (Hulbert [1984]) which is different than our early spring burns and no apparent reductions. Furthermore, two successive years of spring burning in April in a forested corridor of Kansas reduced *S. orbiculatus* canopy cover from 40% to 9% but did not alter shrub species richness (Abrams [1988]). Thus, burning later in the season or burning a site repeatedly for successive years may result in reductions of *S. orbiculatus*. However, at our research sites, most burning is done in the early spring and may not be possible in late spring due to *Schedonorus arundinaceus* (tall fescue) greenup and alteration of fire behavior (McGranahan et al. [2012]). Furthermore, it can be difficult to graze and burn complete pastures for successive years (Ansley et al. [2010]).

Woody plant encroachment in mesic grasslands was regulated by fire (Pyne [1997]; Briggs et al.[2005]). However, fire has largely been removed from the landscape or is only applied in a short seasonal window in the late-dormant season leading to the need for herbicide treatments (Smith[1977]; Defelice [1991]; Stubbendieck et al. [2003]). It is important to consider that historically wildland fire seasonality, intensity and frequency may have been very different than conventional applications of prescribed fire. Our study suggests that a three-year fire return interval using dormant season prescribed fires does not result in plant mortality but can result in altered physical structure and could maintain lower densities of *S. orbiculatus*. While our study did not document mortality, this is a common result for resprouting shrubs (Canadell et al. [1991]; Olson and Platt[1995]). However, we did document lower densities on pastures managed with regular fire compared to higher densities on pastures managed with long-term fire exclusion but the stimulation of aerial layering by fire could theoretically increase density. This contradiction needs additional research across greater temporal scales. Furthermore, unlike other shrub species such as *R. glabra* that may increase with fire, *S. orbiculatus* density and expansion appears to be hindered by regular fire but further research is needed on other seasonal applications and intensities of fire (Hajny et al. [2011]).

Our study also included the spatio-temporal application of patchy fires (PBG) and cattle grazing, an attempt to recouple pre-settlement fire and grazing processes (Burrows [1991]; Fuhlendorf and Engle [2001]). The attraction to recently burned patches alters grazing patterns and results in fuel accumulation in long-unburned patches and potentially enhances fire behavior (Kerby et al.[2007]). However, when compared to attempting to burn a pasture completely, we did not observe different vegetative responses associated with fire intensity or herbivory. This is not surprising because *S. orbiculatus* is an extremely effective resprouter after fire and cattle do not graze *S. orbiculatus* (Stubbendieck et al. [2003]).

CONCLUSIONS

Our documentation of fire triggering aerial layering of *S. orbiculatus* is a new insight relevant to fire-prone mesic grasslands. Other studies have documented physical stimulation of aerial layering for other species in less fire-prone ecosystems or for horticultural purposes (Hartmann and Kester[1975]; O'Dea et al. [1995]; Lantz and Antos [2002]; Deb and Pogener [2012]). As woody plants continue to encroach and transition these critical areas for conservation from grasslands to shrublands, it is increasingly critical that we continue to understand the organismal communities and ecological drivers (Samson [2004]; Briggs et al. [2005]; Zhang and Zhang [2007]). We suggest that aerial layering be considered as an adaptation to fire disturbances and that additional research is warranted. Areas for additional inquiry include how aerial layering and seed production of *S. orbiculatus* fluctuate along a gradient of disturbance and how fire intensity and fire return interval affect belowground total non-structural carbohydrate reserves because modern anthropogenic fire regimes may not replicate a natural lighting-ignited fire regime (Kennedy and Potgieter [2003]; de Groot and Wein [2004]). Finally, it is also important to understand how this basic plant ecology information applies to other resprouting shrub species that use a layering mechanism and how spatially heterogeneous prescribed burning functions in land management and conservation (Hodgkinson [1998]; Doumas and Koprowski [2012]).

AUTHORS' CONTRIBUTIONS

JDS designed the shrub measurement protocol, conducted all sampling and statistical analyses. DME and DMD designed the overarching fire and grazing treatments. RNH served as the fire boss for all prescribed burning activities and established the permanent plots. All authors wrote and revised the manuscript and have read and approved the final manuscript.

ACKNOWLEDGEMENTS

This research was supported by the Oklahoma Agricultural Experiment Station, Iowa Department of Natural Resources (DNR), Iowa Agricultural and Home Economics Experiment Station, and the Iowa State Wildlife Grants program grant #-U-2-R-1 in cooperation with the U.S. Fish and Wildlife Service, Wildlife and Sport Fish Restoration Program (#-U-2-R-1). Appreciation is extended to Josh Rusk and Shannon Rusk for support of field activities and coordination of treatments. We also extend appreciation to the many graduate students, staff and faculty that conducted the 27 prescribed fires over the course of this project.

REFERENCES

1. Abrahamson WG (1975) Reproductive strategies in dewberries. Ecology 56:721-726 doi:10.2307/1935508
2. Abrams MD (1988) Effects of prescribed fire on woody vegetation in a gallery forest understory in northeastern Kansas. Trans Kans Acad Sci 91:63-70
3. Agee JK (1993) Fire ecology of Pacific Northwest Forests. Island Press, Washington DC.
4. Aleksoff KC (1999) Symphoricarpos Oreophilus. In: Fire Effects Information System. US Department of Agriculture, Forest Service, Rocky Mountain Research Station, Fire Sciences Laboratory. http://www.feis-crs.org/beta/ Accessed 30 Dec 2013
5. Anderson ML, Bailey AW (1979) Effect of fire on a *Symphoricarpos occidentalis* shrub community in central Alberta. Can J Bot 57:2819-2823 doi:10.1139/b79-334
6. Ansley RJ, Rasmussen GA (2005) Managing native invasive Juniper species using fire. Weed Technol 19:517-522 doi:10.1614/WT-04-098R1.1
7. Ansley RJ, Pinchak WE, Teague WR, Kramp BA, Jones DL, Barnett K (2010) Integrating grazing and prescribed fire restoration strategies in a mesquite savannah: II. Fire behavior and mesquite landscape cover responses. Range Ecol Manage 63:286-297 doi:10.2111/08-172.1
8. Bartos DL, Mueggler WF (1982) Early succession following clearcutting of aspen communities in Northern Utah. J Range Manage 35:764-768 doi:10.2307/3898260
9. Bellingham PJ, Sparrow AD (2000) Resprouting as a life history strategy in woody plant communities. Oikos 89:409-416 doi:10.1034/j.1600-0706.2000.890224.x
10. Bond WJ, Midgley JJ (2003) The evolutionary ecology of sprouting in woody plants. Int J Plant Sci 164:S103-S114 doi:10.1086/374191
11. Bond WJ, Van Wilgen BW (1996) Fire and plants, Population and Community Biology Series 14. Springer, Netherlands.
12. Bragg TB, Hulbert LC (1974) Woody plant invasion of unburned Kansas bluestem prairie. J Range Manage 29:19-23 doi:10.2307/3897682

13. Brennan LA (1991) How can we reverse the northern bobwhite population decline? Wildlife Soc B 19:544-555
14. Briggs JM, Knapp AK, Blair JM, Heisler JL, Hoch GA, Lett MS, McCarron JK (2005) An ecosystem in transition: causes and consequences of the conversion of mesic grassland to shrubland. Bioscience 55:243-254 doi:10.1641/0006-3568(2005)055[0243:AEITCA]2.0.CO;2
15. Burrows ND (1991) Patch-burning desert nature reserves in Western Australia using aircraft. Int J Wildl Fire 1:49-55
16. Canadell J, Lloret F, López-Soria L (1991) Resprouting vigour of two Mediterranean shrub species after experimental fire treatments. Vegetatio 95:119-126 doi:10.1007/BF00045210
17. Clark JS (1991) Disturbance and tree life history on the shifting mosaic landscape. Ecology 72:1102-1118 doi:10.2307/1940609
18. Collins SL (1992) Fire frequency and community heterogeneity in tallgrass prairie vegetation. Ecology 73:2001-2006 doi:10.2307/1941450
19. Collins SL (2000) Disturbance frequency and community stability in native tallgrass prairie. Amer Nat 155:311-325 doi:10.1086/303326
20. Collins SL, Adams DE (1983) Succession in grasslands. Thirty-two years of change in a central Oklahoma tallgrass prairie. Vegetatio 51:181-190 doi:10.1007/BF00129437
21. de Groot WJ, Wein RW (2004) Effects of fire severity and season of burn on *Betula glandulosa* growth dynamics. Int J Wildland Fire 13:287-295 doi:10.1071/WF03048
22. Deb CR, Pongener A (2012) Studies on the in vitro regenerative competence of aerial roots of two horticultural important *Cymbidium* species. J Plant Biochem Biot 21:235-241 doi:10.1007/s13562-011-0099-5
23. Debinski DM, Moranz RA, Delaney JR, Miller JR, Engle DM, Winkler LB, McGranahan DA, Barney RJ, Trager JC, Stephenson AL, Gillespie MK (2011) A cross-taxonomic comparison of insect responses to grassland management and land-use legacies. Ecosphere 2:1-16 doi:10.1890/ES11-00226.1
24. Defelice MS (1991) Buckbrush (*Symphoricarpos orbiculatus*) control in tall fescue (Festuca arundinacea) pastures. Weed Technol 5:841-844

25. Doumas SL, Koprowski JL (2012) Effect of heterogeneity in burn severity on Mexican fox squirrels following the return of fire. Int J Wildland Fire 22:405-413 doi:10.1071/WF12046
26. Drewa PB (2003) Effects of fire season and intensity on *Prosopis glandulosa* Torr. var.*glandulosa*. Int J Wildland Fire 12:147-157 doi:10.1071/WF02021
27. Enright NJ, Goldblum D (1999) Demography of a non-sprouting and resprouting *Hakea*species (Proteaceae) in fire-prone Eucalyptus woodlands of southeastern Australia in relation to stand age, drought and disease. Plant Ecol 144:71-82 doi:10.1023/A:1009839800864
28. Fuhlendorf SD, Engle DM (2001) Restoring heterogeneity on rangelands: ecosystem management based on evolutionary grazing patterns. Bioscience 51:625-632 doi:10.1641/0006-3568(2001)051[0625:RHOREM]2.0.CO;2
29. Gratani L, Amadori M (1991) Post-fire resprouting of shrubby species in Mediterranean maquis. Vegetatio 96:137-143 doi:10.1007/BF00044975
30. Guthery FS (1980) Bobwhites and brush control. Rangelands 2:202-204
31. Hajny KM, Hartnett DC, Wilson GW (2011) *Rhus glabra* response to season and intensity of fire in tallgrass prairie. Int J Wildland Fire 20:709-720 doi:10.1071/WF09127
32. Harrell WC, Fuhlendorf SD, Bidwell TG (2001) Effects of prescribed fire on sand shinnery oak communities. J Range Manage 54:685-690 doi:10.2307/4003672
33. Hartmann HT, Kester DE (1975) Plant Propagation: Principles and Practices. Prentice-Hall, New Jersey.
34. Hauser AS (2007) Symphoricarpos Occidentalis. In: Fire Effects Information System. US Department of Agriculture, Forest Service, Rocky Mountain Research Station, Fire Sciences Laboratory. http://www.feis-crs.org/beta/ Accessed 30 Dec 2013
35. Hidayati SN, Basakin JM, Baskin CC (2001) Dormancy-breaking and germination requirements for seeds of *Symphoricarpos orbiculatus* (Caprifoliaceae). Am J Bot 88:1444-1451 doi:10.2307/3558452

36. Higgins SI, Pickett ST, Bond WJ (2000) Predicting extinction risks for plants: environmental stochasticity can save declining populations. Trends Ecol Evol 15:516-520 doi:10.1016/S0169-5347(00)01993-5

37. Hodgkinson KC (1998) Sprouting success of shrubs after fire: height-dependent relationships for different strategies. Oecologia 115:64-72 doi:10.1007/s004420050492

38. Holechek JL, Pieper RD, Herbel CH (2001) Range Management: Principles and Practices. Prentice-Hall, New Jersey.

39. Hulbert LC (1984) Fire effects on tallgrass prairie. In: Proceedings of the 9th North American Prairie Conference. Moorehead University, Moorehead.

40. Hullick CJ, Manske LL (2006) Evaluation of Western Snowberry Management Practices: A Study and Demonstration. In: North Dakota State University Archives. http://www.ag.ndsu.edu/archive/dickinso/research/2005/range05f.htm Accessed 30 Apr 2014

41. SAS Institute (2011) SAS data integration studio 4.3: user's guide. SAS institute, North Carolina

42. Iowa Environmental Mesonet (2014) IEM climodat for Mt. Ayr. In: Iowa Mesonet Station, Iowa State University. http://mesonet.agron.iastate.edu/climodat/ Accessed 20 Feb 2014

43. Iwasa YOH, Kubo T (1997) Optimal size of storage for recovery after unpredictable disturbances. Evol Ecol 11:41-65 doi:10.1023/A:1018483429029

44. Kennedy AD, Potgieter ALF (2003) Fire season affects size and architecture of*Colophospermum mopane* in southern African savannas. Plant Ecol 167:179-192 doi:10.1023/A:1023964815201

45. Kerby JD, Fuhlendorf SD, Engle DM (2007) Landscape heterogeneity and fire behavior: scale-dependent feedback between fire and grazing processes. Landscape Ecol 22:507-516 doi:10.1007/s10980-006-9039-5

46. Knapp EE, Schwilk DW, Kane JM, Keeley JE (2006) Role of burning season on initial understory vegetation response to prescribed fire in a mixed conifer forest. Can J Forest Res 37:11-22 doi:10.1139/x06-200

47. Korschgen LJ (1962) Foods of Missouri deer, with some management implications. J Wildlife Manage 26:164-172 doi:10.2307/3798598

48. Kruger LM, Midgley JJ, Cowling RM (1997) Resprouters vs reseeders in South African forest trees; a model based on forest canopy height. Funct Ecol 11:101-105 doi:10.1046/j.1365-2435.1997.00064.x

49. Lantz TC, Antos JA (2002) Clonal expansion in the deciduous understory shrub, devil's club (*Oplopanax horridus*; Araliaceae). Can J Bot 80:1052-1062 doi:10.1139/b02-095

50. McGranahan DA, Engle DM, Fuhlendorf SD, Miller JR, Debinski DM (2012) An invasive cool-season grass complicates prescribed fire management in a native warm-season grassland. Nat Area J 32:208-214 doi:10.3375/043.032.0214

51. McLean A (1969) Fire resistance of forest species as influenced by root systems. J Range Manage 22:120-122 doi:10.2307/3896195

52. McWilliams J (2000) Symphoricarpos albus. In: Fire Effects Information System. US Department of Agriculture, Forest Service, Rocky Mountain Research Station, Fire Sciences Laboratory. http://www.feis-crs.org/beta/ Accessed 30 December 2013

53. McWilliams JD (2005) Symphoricarpos Longiflorus. In: Fire Effects Information System. US Department of Agriculture, Forest Service, Rocky Mountain Research Station, Fire Sciences Laboratory. http://www.feis-crs.org/beta/ Accessed 30 Dec 2013

54. Midgley JJ (1996) Why the world's vegetation is not totally dominated by resprouting plants; because resprouters are shorter than reseeders. Ecography 19:92-95 doi:10.1111/j.1600-0587.1996.tb00159.x

55. Miller WB (1964) An Ecological Study of the Mountain Mahogany Community. Laramie: Thesis, University of Wyoming.

56. Missouri Department of Conservation (2010) Grand River Grasslands, Conservation Opportunity Area. In: Missouri Department of Conservation Wildlife Division. http://mdc.mo.gov/node/9822 Accessed 30 Dec 2013

57. Morgan P, Neuenschwander LF (1988) Seed-bank contributions to regeneration of shrub species after clear-cutting and burning. Can J Bot 66:169-172 doi:10.1139/b88-026

58. Nesmith JC, Hibbs DE, Shatford JP (2006) Clonal growth patterns of creeping snowberry (*Symphoricarpos hesperius* GN Jones) and trailing blackberry (*Rubus ursinus* Cham. and Schlecht) in the western foothills of the Cascade Mountains, Oregon. Northwest Sci 80:274-282
59. NOAA (2014) Climate at a Glance, Time Series. In: National Climatic Data Center, National Oceanic, and Atmospheric Administration. www.ncdc.noaa.gov/cag/ Accessed 29 Sep 2014
60. O'Dea ME, Zasada JC, Tappeiner JC (1995) Vine maple clone growth and reproduction in managed and unmanaged coastal Oregon Douglas-fir forests. Ecol Appl 5:63-73 doi:10.2307/1942052
61. Olson MS, Platt WJ (1995) Effects of habitat and growing season fires on resprouting of shrubs in longleaf pine savannas. Vegetatio 119:101-118 doi:10.1007/BF00045593
62. Pelton J (1953) Studies on the life-history of *Symphoricarpos occidentalis* Hook., in Minnesota. Ecol Monogr 23:17-39 doi:10.2307/1943518
63. USDA Plant Database (2013) Symphoricarpos orbiculatus. In: US Department of Agriculture. https://plants.usda.gov/core/profile?symbol=SYOR Accessed 30 Apr 2014
64. Poreda SF, Wullstein LH (1994) Vegetation recovery following fire in an oakbrush vegetation mosaic. West N Am Nat 54:380-383
65. Pyne SJ (1997) America's Fires. Forest History Society, Durham.
66. Rocca ME (2009) Fine-scale patchiness in fuel load can influence initial post-fire understory composition in a mixed conifer forest, Sequoia National Park, California. Nat Area J 29:126-132 doi:10.3375/043.029.0204
67. Romo JT, Grilz PL, Redmann RE, Driver EA (1993) Standing crop, biomass allocation patterns and soil-plant water relations in *Symphoricarpos occidentalis* Hook. following autumn or spring burning. Am Midl Nat 130:106-115 doi:10.2307/2426279
68. Samson F, Knopf F (1994) Prairie conservation in North America. Bioscience 44:418-421 doi:10.2307/1312365
69. Samson FB, Knopf FL, Ostlie WR (2004) Great Plains ecosystems: past, present, and future. Wildlife Soc B 32:6-15 doi:10.2193/0091-7648(2004)32[6:GPEPPA]2.0.CO;2

70. Scasta JD (2014) Implications of Pyric Herbivory For Grassland Ecology, Management and Production. Stillwater: Dissertation, Oklahoma State University.

71. Shortt KB, Vamosi SM (2012) A review of the biology of the weedy Siberian peashrub*Caragana arborescens*, with an emphasis on its potential effects in North America. Bot Stud 53:1-8

72. Smith L (1977) The Allelopathic Potential of Symphoricarpos orbiculatus in the Kansas Tall Grass Prairie. Emporia: Thesis, Emporia State University.

73. Snyder SA (1991) Symphoricarpos mollis. In: Fire Effects Information System. US Department of Agriculture, Forest Service, Rocky Mountain Research Station, Fire Sciences Laboratory. http://www.feis-crs.org/beta/ Accessed 30 Dec 2013

74. Soper RB, Lochmiller RL, Leslie DM, Engle DM (1993) Nutritional quality of browse after brush management on cross timbers rangeland. J Range Manage 46:399-410 doi:10.2307/4002657

75. Stubbendieck JL, Coffin MJ, Landholt LM (2003) Weeds of the Great Plains. Nebraska Department of Agriculture Bureau of Plant Industry, Nebraska.

76. Theis N, Donoghue MJ, Li J (2008) Phylogenetics of the Caprifolieae and Lonicera (Dipscales) based on nuclear and chloroplast DNA sequences. Syst Bot 33(4):776-783

77. USDA NRCS (2013) Iowa soil regions. In: US Department of Agriculture, Natural Resource Conservation Service. http://www.nrcs.usda.gov/Internet/FSE_DOCUMENTS/nrcs142p2_006361.pdf Accessed 30 Apr 2014

78. Wang SYS, Barandiaran D, Hilburn K, Houser P, Oglesby B, Pan M, Pinker R, Santanello J, Schubert S, Wang H, Gillies R (2014) Could the 2012 drought in the central US have been anticipated? A review of the NASA working group research. J Earth Sci Eng 4:428-437

79. Wells PV (1969) The relation between mode of reproduction and extent of speciation in woody genera of the California chaparral. Evolution 23:264-267 doi:10.2307/2406790

80. Westman WE, O'Leary JF (1986) Measures of resilience: the response of coastal sage scrub to fire. Vegetatio 65:179-189 doi:10.1007/BF00044818

81. Winter SL, Fuhlendorf SD, Goad CL, Davis CA, Hickman KR, Leslie DM (2011) Fire tolerance of a resprouting *Artemisia* (Asteraceae) shrub. Plant Ecol 212:2085-2094 doi:10.1007/s11258-011-9975-2
82. Youngblood A, Metlen KL, Coe K (2006) Changes in stand structure and composition after restoration treatments in low elevation dry forests of northeastern Oregon. Forest Ecol Manag 234:143-163 doi:10.1016/j.foreco.2006.06.033
83. Zhang JT, Zhang F (2007) Diversity and composition of plant functional groups in mountain forests of the Lishan Nature Reserve, North China. Bot Stud 48:339-348

Chapter 9

Nuisance Species: Beyond the Ecological Perspective

Zina Skandrani[1], Sébastien Lepetz[2], and Anne-Caroline Prévot-Julliard[1]

[1]Centre d'Ecologie et des Sciences de la Conservation CESCO, UMR 7204 CNRS-MNHN-UPMC, Museum National d'Histoire Naturelle, CP 51, 55 rue Buffon, F-75005 Paris, France

[2]Laboratoire d'Archéozoologie et Archéobotanique, UMR 7209 CNRS-MNHN, Museum National d'Histoire Naturelle, CP 51, 55 rue Buffon, F-75005 Paris, France

ABSTRACT

Introduction

'Nuisance species' (or 'invasive' species) are often proposed to be the second most important concern in the context of the current biodiversity crisis. Despite increasing evidence that exotic species do not always become invasive, this perception is still common in the scientific community. This suggests that other issues are at stake in the concept of nuisance species rather than just the problems they raise, grounded mostly in ecological or economic arguments.

Methods

We retraced the evolution of pigeon representation in France through an extensive review of ancient texts related to pigeons and reviewed more than 240 source texts, dating from the seventh century to the 1990s. We completed these data with literature on human-animal relationships and animal conceptions in religious, philosophical, scientific, and political currents of thought that were representative of Western and, specifically, French tradition.

Results

We used the heated debate over pigeons (*Columba livia*) as pests as a case study by analyzing the ecological, social, economic, and political relations regarding pigeons. Through a historical perspective of the debate about and understanding of pigeons, we propose a new and complementary explanation for the modern thinking of pigeons as a nuisance species, based on what we termed their socio-nature characteristics. In particular, we used social representations theory to highlight the issues of human identity construction at stake in the construction of pigeons as a nuisance species.

Conclusions

We invite the reader to consider the impact of such human self-definition on environmental stances, as it could constrain further developments or improvements of conservation perspectives.

INTRODUCTION

'Nuisance species' (or 'invasive species' or 'alien species') are often proposed to be the second most important concern in the context of the current biodiversity crisis (Lowe *et al.* 2000; Wilcove *et al.*1998) because they are a potential threat to native species through predation, grazing, competition, parasitism, disease, hybridization, or habitat alteration (Nunez *et al.* 2010; Elliott *et al.* 2010; Salo *et al.* 2006; Mooney *et al.* 2005; Pimentel *et al.* 2001). Since Elton (1958), the bias against alien species, which was expressed in their description as nuisances, has widely been embraced by the public, scientists, conservationists, land managers, and policy-makers and has given birth in the 1990s to a discipline in its own right, 'invasion biology'. In biological conservation textbooks, invasive or nuisance species are described as more dangerous than environmental degradation, habitat fragmentation, and pollution because these may be more restrained and reversible than the reproduction and dispersal dynamics of nuisance species (for example, Primack *et al.* 2012). In addition to their potential devastating effects on natural ecosystems, nuisance species also have more general negative impacts on human beings. The IUCN, for instance, has proposed: 'They are causing significant damage to ecological, economic, and health levels. As a matter of fact, they can compete with native species, act as pests or pathogens for cultivated or domesticated species, or even disseminate allergic or infectious agents.' (International Union for Conservation of Nature, 2013).

These generalized discourses about invasive species, however, have been increasingly debated and challenged in recent years (Marion 2013; Davis *et al.* 2011; Valéry *et al.* 2008; Warren 2007; Falk-Peterson *et al.* 2006; Colautti and MacIsaac 2004) than other causes of biodiversity loss (Stuart *et al.*2004; Davis *et al.* 2011). Although some alien species can constitute real threats to native species or habitats, case-by-case studies are needed to explore their context-dependent impacts. Some species introduced by human beings are indeed destructive, for instance, zebra mussels (*Dreissena polymorpha*) in the Great Lakes, while in other places they can increase biodiversity (the honeysuckle, for instance, favors native bird species in Pennsylvania; Davis *et al.* 2011). It is also possible for native species to have similar devastating effects (Valéry *et al.* 2009). Because it is impossible to return to past ecosystems that have been increasingly transformed by a plethora of irreversible drivers of change (climate change, nitrogen eutrophication, increased urbanization, and other land-use changes), there has been an increased push to acknowledge 'novel ecosystems' (Chapin and Starfield 1997), which are composed of new combinations of species under new abiotic conditions. Therefore, many authors advocate for adaptive ecosystem management, focusing on the resilience of desirable states, rather than targeting the often-impossible eradication of undesirable species (for example, Seastedt *et al.* 2008).

Despite evidence that the nuisance status of some species has often been globally exaggerated, this 'bioxenophobia' (Marion 2013), which is based on the myth of stable and pure fauna or flora, persists. This suggests that more issues are at stake regarding the concept of nuisance species than the explicit ecological and environmental problems (Fazey *et al.* 2006; Wallace 2003). By widening our perspective, we can recognize that environments are combined constructions of physical, ecological, and social elements (Swyngedouw 1999). In this perspective, socio-physical constructions and environmental issues, such as nuisance species, are the consequences of interwoven socio-ecological processes (Heynen 2006). Following Swyngedouw (1999:445), 'natural or ecological conditions and processes do not operate separately from social processes, and [...] the actually existing socio-natural conditions are always the result of intricate transformations of pre-existing configurations that are themselves inherently natural and social.' (See also, Minteer and Collins2005).

In this paper, we provide non-ecological evidence that advocates for the need to use caution when dealing with the alien species issue. We argue for a novel explanation of nuisance species as belonging to 'socio-natures' (see Swyngedowu 2004,1999, Castree and Braun 2001) that, beyond *de facto* ecological considerations, are historically produced and continuously reformed by networks of human and non-human 'actants' (Latour 1993). 'In such a network of associations, each of the separate pieces is not independent, but is instead made to be the way it is by virtue of its relationship to all the other parts. [...] Nor is the identity of each actor independently formed and then joined to the whole. Instead, each becomes what it is through its specific relation to the other', (Robbins2007:14).

We used the case study of the heated debate over feral pigeons (*Columba livia*), which are widely considered pests, to explore the socio-nature of nuisance species (Blechman 2006). We propose a new approach to this problem through the analysis of ecological, social, economic, and political relations, as well as power geometries, underlying this problem (Heynen 2006; Blaikie 1999). The case of the pigeon is particularly telling in this regard, as they have over history 'perhaps taken on more symbolic and functional roles than any other bird,' (Jerolmack 2008:74). Indeed, they originate from formerly domesticated pigeons that returned to a free state of living (that is, ferality). These were the first birds to be domesticated over 5,000 years ago, as their ancestors, wild rock doves, came to forage on cultivated fields in direct proximity to human beings (Johnston and Janiga 1995). Domestication created the white-colored pigeons that are famous in mythology, religions, and literature as 'doves'. Feral urban pigeons (which we shall refer to as urban pigeons) are found in all major cities around the world (del Hoyo *et al.* 2005) and are today the birds with the closest contact to people in cities; currently, these pigeons are mostly ascribed with a highly negative image (Vuorisalo *et al.* 2001).

Relying on an examination of the changing meaning of pigeons, we clarified, in some historical depth, what particular discourses and understandings of and about non-human beings are mobilized in the modern perception of pigeons as a nuisance species. Finally, we show that pigeons themselves have their own agency in shaping these (re)-configurations and hold an essential and active part in the system.

METHODS

We retraced the evolution of pigeon representation in France through an extensive review of ancient texts related to pigeons in the following domains: religious, legal, economic, administrative, and literary, and in military reports, and treatises on zoology, agriculture, medicine, and cooking. We used texts available from the French numerical library *Gallica* from the Bibliothèque Nationale de France (2013), which provides access to over a million and a half documents, representative of the documents issued in France over the past centuries, that have been digitized by the Bibliothèque Nationale de France and other partner libraries. We searched for the French terms for pigeon (*pigeon*) and dove (*colombe*) and reviewed more than 240 source texts, dating from the seventh century (except for the Old Testament, which dates from the eighth to the second century BCE) to the 1990s. We narrowed the selection to 50 texts that reflected the variability of all the information and arguments. We selected these texts based on their pertinence to our subject, as a large part of the reviewed documents mentioned the words pigeon (*pigeon*) and dove (*colombe*) without providing any relevant information or providing repetitive material, such as the dove being a representation of the Holy Spirit.

To put pigeon representations into their historical frameworks, we completed these data with literature on human-animal relationships and animal conceptions in religious, philosophical, scientific, and political currents of thought that were representative of Western and, specifically, French tradition.

RESULTS

Based on our historical search, we found a strong shift in the mid-twentieth century in the image of the pigeon, from a very positive historical representation to its current mostly negative image in mainstream public opinion.

When Pigeons Were Still Gentle Doves...

In the religious realm, doves were often portrayed as animal companions to gods, such as the love goddess Kamadeva in Hinduism, and also as representations of gods themselves, as with Aphrodite and Venus, the goddesses of love in Ancient Greece and Rome (La Ville de Mirmont 1894). In Christianity, Christ received the Holy Spirit descending in the shape of a dove during his baptism (Raban Maur 842–847, in Voisenet 2000). In Islam, a pair of pigeons saved the prophet Mohamed's life by nesting in front of the cave he was hiding in and, in this way, misled his persecutors (Lings1986).

Doves hold the role of divine messengers in various cultures. In Ancient Egypt, they uttered to oracles (Allys 1909, Bonnery 1909); in Ancient Rome, they constituted divine auguries (La Ville de Mirmont 1894); in the Old Testament story of Noah's Ark, they announced that the waters of the Flood had drained away by bringing back an olive branch from terra firma and thus announced God's forgiveness (Jerolmack 2007); and finally, in the Christian Hagiography they revealed to a pregnant mother the sanctity of her child by laying three drops of milk on her mouth, and afterwards the saint's impending martyrdom, through three drops of blood (Raban Maur 842–847, in Voisenet2000). Consequently, the dove is the most cited animal within the Holy Scriptures (it is cited nearly 50 times) and illustrates Christian virtues, such as purity, simplicity of heart and soul, gentleness, and chastity (Voisenet 2000). These sacred representations largely influenced the famous symbol of the dove of peace and the literary figure of the romantic dove (La Fontaine 1678), that are still common in modern times.

The attribution of these symbolic charges to doves was related to the functions of pigeons in historical times. In medieval Christian logic, animals were symbolically separated into two opposite groups, representing good or Christ and the devil or Satan, to serve a moral purpose. These representations were supposed to help human beings situate themselves between vice and virtue and choose one of the two orders (Ribémont 1992). Birds, and among them doves, were used in this context as a pedagogical model to convey a spiritual message, encouraging human beings to behave better in life. Based on their faculty to move between the sky and the earth and to escape terrestrial confines, birds, and doves, established a connection between reality and the ideal, as well as the material and the spiritual. By linking them, they showed that conciliation between both was possible (Ribémont1992). The wing symbolism was largely developed to exemplify the elevation of the human spirit and soul. It illustrated the progress dynamic and invited human beings to surpass themselves. Furthermore, because of their monogamous habits (Giunchi *et al.* 2007), doves also incarnated the ideal of the morally valued chastity and monogamy. Thus, doves became an instrument of the divine and were positively charged in support of the Christian quest for perfection and redemption (Voisenet, 2000).

However, in addition to this moral function, domestic pigeons also had different uses in everyday life. They were bred for their meat and for the use of their nitrogen-rich feces as fertilizer (Thiébaut de Berneaud 1841, Bourguignon 1882), to be used as messengers for communication purposes, and (although to a lesser extent) they were used for pleasure or sport (Johnston and Janiga 1995; de Planhol 2004). In the first French encyclopedias, which related the customs and representations of their time, pigeons were described as extremely profitable animals, and a special emphasis was placed on their value through the listing of their numerous uses and different breeds (Corneille1694; Diderot and D'Alembert 1751; Lunel 1858).

Pigeon meat was, until the second half of the twentieth century, the second most eaten poultry and was mentioned just after chicken in cookbooks (Menon 1749; Foy 1845; Viard 1851; Courchamps1853; Breteuil 1860) and treatises on agriculture (Rengade 1887; Brévans 1892; Favre 1905; Bonnier 1922). In France, the right to have dovecotes and keep pigeons was reserved for lords (De Ferriere 1741; Renauldon 1765) until the French Revolution in 1789. Pigeons were thus, in this context, an outward sign of nobility (Hoüard 1780; Henry 1981). After the abolition of this aristocratic privilege in 1789 (Gayot 1876; Rupin 1884), the breeding of pigeons was encouraged in the countryside and in cities as a means to provide families of the lower working class with meat and an additional financial resource until World War II (Bois 1847). Pigeon meat was indeed regularly cited in recipes in the gastronomic sections of journals during that period and in serial narrations of feasts. However, it appears from the sophisticated character of these menus that pigeon meat was still considered a delicacy and had maintained its former noble connotations inherited from its status under the Ancien Régime (Givois, unpublished data). The use of pigeon organs, blood, and feces or as a whole in popular medicine (Moquin-Tandon 1860; Spalikowski 1895; Reutter 1916; Hermant and Boomans 1928), contributed to this image of a precious and valuable animal.

The highest praise for pigeons was, however, expressed in the aftermath of the support they provided as messengers (they can be bred as carrier or homing pigeons) during various military conflicts. Pigeons were indeed already used for postal services in ancient times by Egyptians, Greeks, Romans, Hebrews, Persians, Chinese and Arabs, they were (re-)imported to early medieval Europe during the Crusades and in the context of commercial exchange (Van der Linden 1950). They were thus employed at the time of Caesar's conquest of Gaul to inform Rome of the progress of the invasion, during Hannibal's advancement through Europe (Jerolmack 2007) to announce the loss of the Battle of Waterloo to Napoleon I (Palliez 1932), or in a more civil context, to report earthquakes in Japan, to announce the return of ships of all nationalities into harbors (Gérardin 1905), and for communication between distant castles (Guidez 1969). In the nineteenth and twentieth centuries, messenger or soldier pigeons were used at an even larger scale during the Franco-Prussian War, and the liberation of besieged Paris in 1871 was only made possible through the exchange of thousands of messages and photographs carried by pigeons between Paris and London (Bulletin Technique du Génie Militaire 1936). Consequently, the breeding of homing pigeons was publicly advocated by political leaders and scientific experts. Despite the coming age of modern communication technologies, such as the telegraph and telephone, pigeons remained crucial during World Wars I and II to request munitions military support on the front lines and to inform about the progress of the enemy lines, as the other communication lines were easily destroyed and messages were intercepted or tapped (Gérardin 1905; Palliez 1932; Jerolmack 2007).

The ensuing allocation of public funding by the French government for pigeon fancier associations (Conseil Général. Rapports et Délibérations 1893, 1894, 1902, 1903, 1926, 1934) and the erection of monuments dedicated to pigeon soldiers in Lille (France) and Brussels (Belgium) confirm the extremely positive representation of pigeons during this period. At the same time, feeding street pigeons was such a common leisure pastime that urban parks had dedicated feeding sites and seed sellers.

Pigeons (Un) Wanted: Dead or Alive

This image is in sharp contrast with the current often-expressed dislike for pigeons. The progressive transformation of the representation of pigeons, from positive to negative, in the mid-twentieth century occurred contemporaneously with their unchecked proliferation in cities (Vuorisalo 2001), sanitary considerations, and the disappearance of historical uses for pigeons. The presence of pigeons in cities was already mentioned, anecdotally, in Nuremberg (Germany) in the twelfth century and in London (United Kingdom) in the fourteenth century (de Planhol 2004). In France, the pigeon was described in treatises and natural histories of pigeons at the end of the nineteenth century and at the beginning of the twentieth century as a free-ranging bird of domestic origins, living on farms and in the countryside (La Perre de Roo 1877; Barot 1891; Valette 1910). Pigeons were further counted among migrant species (Gayot 1876) and reported to be present only two or three times a year (Crespon 1844). From the mid-twentieth century on, pigeons moved towards urban areas and into human proximity. It is thus a consequence of a major change in pigeon ecology that they took over major Western cities. More precisely, the growth of urban pigeon populations through the colonization of a novel environment could be viewed as the result of an 'ecological fitting' (Agosta and Klemens 2008, 258 Janzen 1985) process, favored and triggered by various favorable factors: a large amount of food on city streets and marketplaces, numerous nesting sites in buildings, increased farming productivity with surpluses of grain (Johnston and Janiga 1995), and the high annual reproduction rate of the pigeons that had been selected during their former domestication. Important urban populations of pigeons developed all over Europe during this period (for example, in Hamburg, where the number grew from three nests in 1901 to 3,000 individuals in 1938 and 20,000 in 1963). In the 1960s, their numbers were estimated to be 100,000 in Munich (Germany), 60,000 in Turin (Italy), 200,000 in Vienna (Austria), and 8,000 to 10,000 in Nuremberg (Germany); the same tendency also occurred in the United States (de Planhol 2004). However, quantitative data on pigeon numbers were not always available and led to frightening speculations. For instance, during a city council meeting in Paris in 1969, the pigeon population was estimated to be one million (Bulletin Officiel Municipal de la ville de Paris 1969) without any census (Baud and Brugère 1995); this estimation has since proven to have been excessively exaggerated (Vacher 1970).

Claims about the nuisance caused by pigeon droppings in cities began to be expressed as early as the 1930s and 1940s in journal articles in Europe and overseas (Givois, unpublished data; Jerolmack2008). Complaints that pigeon feces were dirtying and harming historical buildings and monuments became common in French newspaper articles in the following decade (Givois, unpublished data). The ambiguous discourse between the promotion of pigeon breeding on the one hand and the critics on the other hand reveals the growing division occurring between captive domestic pigeons and feral urban pigeons. However, no public measures against pigeons were undertaken before the 1950s, most likely because of their status as a human auxiliary: it was previously a criminal offense to harm pigeons as they could be someone's property (Renauldon 1765; Charriaut 1870). However, in 1955, urban pigeons were legally declared *res nullius*, that is, ownerless, belonging neither to the city of Paris nor to any local authority (Bulletin Officiel Municipal de la ville de Paris 1955). Public opinion regarding pigeons encountered a turning point when they became associated not only with overcrowding, defecating and dirtying monuments, statues and buildings but also with carrying viruses and diseases. In particular, the discovery of the bacterium *Chlamydia psittaci* (Eb 1993), which can potentially cause pulmonary problems, was relayed by a 1952 article that claimed that 70% of pigeons in Paris were carrying the disease (Lépine 1952). These scientific data raised debates about pigeons as a threat to humans, comparing pigeons to 'rats with wings' (Jerolmack 2008).

In France, feeding of urban pigeons in the outer realm became prohibited by law in 1966 (Vacher1970), more or less at the same period as in other Western cities. At the same time, urban pigeon populations started to be controlled by trapping, poisoning, shooting, gassing, and so on (Haag-Wackernagel, 2002; Jerolmack 2008).

These anti-pigeon measures occurred during the emergence of a wide-reaching, hygienist conception of the modern city in the twentieth century in Europe and the United States (Hodak 1999; Leblanc2003), that is, 'orderly and sanitized, with nature subdued and compartmentalized' (Jerolmack2008:72). This hygienism was explained as a reaction to the massive arrival of animals and to ever-growing cities with proportionately increased city dweller requirements in terms of consumption of animal products at the end of the nineteenth century. Entire herds of animals were brought to the city daily, crammed together in the streets, to be slaughtered and quartered, often in open sheds. The exhibition of bloody carcasses on the butchers' façades; the boom of tanneries; the multiplication of horses, dogs, and cats, and thus, the omnipresence of manure, excrement, and urine; the development of maggots; smells of industrial processing of animal products; and the violent sight of animal mistreatment resulted in the creation and intensification of an olfactory and sonic world (Baratay 2009). The unceasing coming and going of animals provoked chaotic traffic, especially on market days, and exhausted animals and hastily trained drivers caused increasingly more incidents. Poultry and rabbits were raised in humble families' households, whereas cows and goats, which were kept in the numerous urban barns, wandered around daily to provide people with fresh milk (Baratay, 2009). Complaints, increased conflicts between animal users and others in their neighborhood, as well as growing attention to sanitary and hygienist concerns forced municipalities to intervene (Baratay, 1998). Programs of urban renovation were launched at the end of the nineteenth century, which cleaned up the city and pushed slaughterhouses to the fringes. This redevelopment of the urban landscape not only included larger streets and squares but also vegetation, instead of animals, as a structuring element of urban space, as well in an aesthetic aim to clean up the air (Leblanc 2003).

In parallel, pigeons held no more functional utility in the beginning of the twentieth century, after their breeding was, to a great extent, abandoned. Pigeon fertilizer was supplanted by the expansion of chemical fertilizer, which began at the end of the nineteenth century (Déhérain 1892; Lagrue1897) and increased after World War II with the modernization and the development of large-scale agriculture (Prévost 2006). The annual delivery of pigeon meat decreased in France from 209 to 67 tons between 1973 and 1979, in favor of the extensively grown chicken, because larger fillets were requested (Royer 1983). The use of pigeons as army messengers also became obsolete after the end of World War II. Pigeons did not even have a function as a moral guide when the Christian religion progressively abandoned its symbolic bestiary from 1930 to 1950 as a consequence of its internal evolution. Animals have indeed been stripped of their religious functions in an attempt to break with rural connotations and to humanize, rationalize and adapt religion to urban life (Baratay 1994).

However, new behavioral particularities observed in urban pigeons could have induced the erosion of their former moral function. First, as with other Columbidae, pigeons feed their young with crop milk and they can reproduce throughout the year as soon as adults can obtain enough food (Johnston and Janiga 1995). Thus, urban pigeons can reproduce throughout the year (with lower reproduction in winter, Jacquin *et al.*2010), which led to them becoming a symbol of hypersexuality and debauchery rather than chastity or other moral values, in a secular society. Second, the alimentary behavior of urban pigeons changed from seed-eating behavior to an omnivorous diet (Johnston and Janiga1995): although they originally fed on seeds found in the fields (Richelet 1680; Diderot and D'Alembert 1751), they switched to foraging on society's waste in the urban context. By continuously walking on the city ground, pigeons stand in opposition to the upstroke of birds and so have broken their symbol of elevation. In a French doctoral thesis in 1970, they were even labeled as 'pigeons de sol', literally 'ground pigeons' (Vacher 1970). As scavengers, pigeons became the incarnation of the ordinary, if not mediocrity, in contradiction to their supernatural aspect (bright whiteness, extraordinary behavior, and sudden appearance) and their higher condition in religious imaginary.

CONCLUSIONS

In each historical period, pigeons were as much a vehicle for the creation and maintenance of social systems as they were a product of those systems. As soon as they stopped complying with the normative network dynamics and the dominant paradigm of valorizing or depreciating animals based on their function and subordination towards humans, they were themselves reformed as a nuisance species for the purpose of human identity construction. Through the pigeon example, we have shown that the concept of nuisance species relies on complex dynamics that, while certainly and definitely including the animals' ecological and behavioral characteristics (here, their abundance and foraging behavior in cities), also involves interactions with socio-political processes. This last aspect remains widely ignored when dealing with nuisance species in conservation issues. However, considering nuisance species as a type of socio-nature further allows turning the problem around; rather than a question of nuisance species, the current failure of the modern society to stop the biodiversity crisis despite such a high level of knowledge and power could be seen as being precisely grounded in the Western paradigm of environmental dominance and exploitation. Indeed, 'still today, terms such as "sustainable development" and "natural resource management" imply the imposition of human values on and control over living nature' (O'Rourke 2000: 147). In that case, (scientific) authorities should be conscious of and reflective on their role and impact and lastly their responsibility to the development and structure of present-day environmental stances.

AUTHORS' CONTRIBUTIONS

ZS carried out the literature survey and wrote the manuscript. ACPJ and SP helped to draft the manuscript. All authors read and approved the final manuscript.

ACKNOWLEDGEMENTS

We would like to thank Professor Eric Baratay and Dr. Assaf Shwartz for helpful discussions and suggestions about this article, as well as two anonymous referees for their comments.

REFERENCES

1. Agosta SJ, Klemens JA (2008) Ecological fitting by phenotypically flexible genotypes: implications for species associations, community assembly, and evolution. Ecol Letters 11:1123-1134
2. Allys (1909) Le Pigeon dans l'Egypte Ancienne. La Correspondance d'Orient 12/1909:1053-1054
3. Andreouli E (2010) Identity, positioning and self-other relations. Papers on Social Representations 19:14.1-14.13
4. Baratay E (1994) Les Controverses Contemporaines sur le Statut de l'Animal. L'Exemple de l'Eglise Catholique, France, 1940–1990. Revue d'histoire moderne et contemporaine 41(3):499-514
5. Baratay E (1995) Zoologie et Eglise catholique dans la France du XVIIIe siècle (1670–1840): une science au service de Dieu. Revue d'histoire des sciences 48(3):241-266.
6. Baratay E (1997) Un Instrument Symbolique de la Domestication: Le Jardin Zoologique aux XIXe-XXe Siècles (L'Exemple du Parc de la Tête d'Or à Lyon). Cahiers d›Histoire 3–4, L›animal domestique. 677-706
7. Baratay E (1998) L'Anthropocentrisme du Christianisme Occidental, In: B. Cyrulnik (ed.), Si les lions pouvaient parler. Essais sur la condition animale, Gallimard, Paris. pp 1428-1459
8. Baratay E (2009) Le Grand Rapprochement. L'Histoire 338:78-87
9. Baratay E (2010) Penser l'animal dans le catholicisme français contemporain (1940–2010). Klesis Revue philosophique – Humanité et animalité 16:201-2012
10. Barot A (1891) Leçons de Choses. Alcan, Paris.
11. Baud C, Brugère H (1995) Chiens et Pigeons en Milieu Urbain: Population, Pollutions, Solutions. Thèse Méd. Vét. Alfort, Paris.

12. Bibliothèque Nationale de France (2013) Gallica bibliothèque numérique. http://gallica.bnf.fr/ Accessed 1 Nov 2011
13. Blaikie P (1999) A review of political ecology. issues, epistemology and analytical narratives. Zeitschrift für Wirtschaftsgeographie Jg 43(3–4):131-147
14. Blechman AD (2006) Pigeons. The fascinating saga of the world's most revered and reviled bird. Grove Press, New York.
15. Bois M (1847) Le Nouvel Art d'Elever et de Multiplier les Pigeons. Tissot, Paris.
16. Bonnery G (1909) Pages Oubliées, Légendes et Traditions. Monnoyer, Le Mans.
17. Bonnier G (1922) Technologie: zoologie appliquée. Les Fils d'Emile Deyrolle Editeur, Paris.
18. Bourguignon A (1882) Volontariat d'un An. Agriculture. Garnier Frères, Paris.
19. Breakwell GM (1993) Social representations and social identity. Papers on Social Representations 2(3):1-20
20. Breteuil J (1860) Le Cuisinier Européen. Garnier Frères, Paris.
21. Brévans J (1892) Le Pain et la viande. Baillère, Paris.
22. Bulletin Officiel Municipal de la ville de Paris (1955) Paris.
23. Bulletin Officiel Municipal de la ville de Paris (1969) Paris.
24. Bulletin Technique du Génie Militaire (1936) Paris.
25. Castree N, Braun B (2001) Social nature. Blackwell, New York.
26. Chapin FS III, Starfield AM (1997) Time lags and novel ecosystems in response to transient climate change in Arctic Alaska. Clim Change 35:449-461
27. Charriaut F (1870) Collection Générale des Lois et Décrets du Gouvernement. Gouvernement de la Défense Nationale, Paris.
28. Clutton-Brock J (1994) The unnatural world: behavioural aspects of humans and animals in the process of domestication. In: Manning A, Serpell JA (eds) Animals and human society: changing perspectives, James Serpell, London. pp 23-35
29. Cohen-Scali V, Moliner P (2008) Représentations Sociales et Identité: Des Relations Complexes et Multiples. L'orientation scolaire et professionnelle 37(4):465-482

30. Colautti RI, MacIsaac HJ (2004) A neutral terminology to define 'invasive' species. Divers Distrib 10:135-141.
31. Conseil Général. Rapports et Délibérations (1893) Marne.
32. Conseil Général. Rapports et Délibérations (1894) Marne.
33. Conseil Général. Rapports et Délibérations (1902) Eure.
34. Conseil Général. Rapports et Délibérations (1903) Marne.
35. Conseil Général. Rapports et Délibérations (1926) Morbihan.
36. Conseil Général. Rapports et Délibérations (1934) Marne.
37. Corneille (1694) Pigeon. Dictionnaires des 16è et 17è Siècles. Huguet, Paris.
38. Courchamps M (1853) Dictionnaire Général de la Cuisine Française Ancienne et Moderne ainsi que de l'Office et de la Pharmacie Domestique. Plon Frères, Paris.
39. Crabtree PJ (1993) Source early animal domestication in the Middle East and Europe. Archaeol Method Theory 5:201-245
40. Crespon J (1844) Faune méridionale ou Description de Tous les Animaux Vertébrés Vivans et Fossiles, Sauvages ou Domestiques [...] du Midi de la France. Editions Camariguo, Nîmes.
41. Davis MA, Chew MK, Hobbs RJ, Lugo AE, Ewel JJ, Vermeij GJ, Brown JH, Rosenzweig ML, Gardener MR, Carroll SP, Thompson K, Pickett STA, Stromberg JC, Del Tredici P, Suding KN, Ehrenfeld JG, Grime JP, Mascaro J, Briggs JC (2011) Don't judge species on their origins. Nature 474:153-154PubMed Abstract |.
42. De Ferriere C (1741) Nouveau Commentaire sur la Coutume de la Prevoté et Vicomté de Paris. Chez Saugrin, Paris.
43. de Planhol X (2004) Le Paysage Animal. L'Homme et la Grande Faune: Une Zoogéographie Historique. Fayard, Paris.
44. Déhérain PP (1892) Amendements et Engrais: Engrais chimiques. In: Masson G (ed) Traité de Chimie Agricole: Développement des Végétaux, Terre Arable, Amendements et Engrais, Corbeil, Paris. pp 792-817
45. del Hoyo J, Elliott A, Christie D (eds) (2005) Handbook of the birds of the world vol. 10 Lynx Ediciones, Barcelona.
46. Delort R (1984) Les Animaux ont une Histoire. Editions du Seuil, Paris.

47. Descola P (2004) Le sauvage et le domestique. Communications 76:16-39

48. Diderot D, D'Alembert J-R (1751) Pigeon. Encyclopédie ou Dictionnaire Raisonné des Sciences, des Arts et des Métiers. Le Breton Editeur, Paris.

49. Digard J-P (1988) Jalons pour une anthropologie de la domestication animale. domestication et représentations, L'Homme 108, Les Animaux. pp 27-58

50. Digard J-P (1990) L'Homme et les Animaux Domestiques. Anthropologie d'une Passion, Fayard, Paris.

51. Digard JP (1992) Perspectives anthropologiques sur la relation homme-animal domestique et sur son évolution. In: Durand R (ed) L›homme, l›animal domestique et l›environnement du Moyen Âge au XVIIIe siècle, Paris. pp 21-27

52. Douglas M (1966) Purity and danger. Penguin, London.

53. Driscoll CA, Macdonald DW, O'Brien SJ (2009) From wild animals to domestic pets, an evolutionary view of domestication. Proc Natl Acad Sci USA 106(1):9971-9978

54. Ducos P (1978) Domestication defined and methodological approaches to its recognition in faunal assemblages. In: Meadow RH, Zeder MA (eds) Approaches to faunal analysis in the Middle East, Peabody Museum Bulletins No. 2, Cambridge. pp 53-56

55. Duveen G (2001) Representations, Identities, Resistance. In: Deaux K, Philoguène G (eds) Representations of the Social, Blackwell, Oxford. pp 257-270

56. Eb F (1993) Chlamydia psittaci: Epidémiologie, pouvoir pathogène et diagnostic. Médecine et Maladies Infectieuses 23(12):919-924

57. Elliott GP, Wilson PR, Taylor RH, Beggs JR (2010) Declines in common, widespread native birds in a mature temperate forest. Biol Cons 143:2119-2126.

58. Elton C (1958) The ecology of invasions by animals and plant. Methuen, London.

59. Falk-Peterson J, Bohn T, Sandlund OT (2006) On the numerous concepts in invasion biology. Biol Inv 8:1409-1424

60. Favre J (1905) Dictionnaire Universel de Cuisine Pratique. Encyclopédie Illustrée d'Hygiène Alimentaire: Modification de l'Homme par l'Alimentation. Librairie Imprimerie des Halles et de la Bourse de Commerce, Paris.

61. Fazey I, Fazey JA, Salisbury JG, Lindenmayer DB, Dovers S (2006) The nature and role of experiential knowledge for environmental conservation. Environ Conserv 33(1):1-10.
62. Foy F (1845) Manuel d'Hygiène. Hachette, Paris.
63. Gayot E (1876) Le Pigeon, Histoire Naturelle, Races d'Utilité et d'Amateurs, Reproduction, Education, Hygiène, Maladies. Tomlin Press, Paris, Ecoles de Tir.
64. Gérardin L (1905) Le Pigeon Messager dit Voyageur au XXe Siècle. O. Bornemann, Paris.
65. Giunchi D, Baldaccini NE, Soldatini C (2007) On the use of pharmacological sterilisation to control feral pigeon populations. Wildlife Res 34:306-318
66. Guidez J (1969) Le Pigeon Voyageur. Bibliothèque de Travail 691:4-34
67. Haag-Wackernagel D (2002) Feral pigeons: management experiences in Europe, in Atti 2nd Convegno Nazionale sulla Fauna Urbana 'Specie ornitiche problematiche: biologia e gestiona nelle città e nel territorio', Firenze, 10 giugno 2000, ARSIA e LIPU. Firenze. M. Dinetti, Regiona Toscana. pp 25-37
68. Henry Y (1981) Le colombier, un Signe Extérieur de Noblesse. Essai sur les Colombiers en Bretagne. Annales de Bretagne et des pays de l'Ouest 88(1):67-86.
69. Hermant P, Boomans D (1928) La Médecine Populaire. 43-45
70. Heynen N, Kaika M, Swyngedouw E (2006) Urban political ecology. Politicising the production of urban natures. In: Heynen N, Kaika M, Swyngedouw E (eds) In the nature of cities: urban political ecology and the politics of urban metabolism, Routledge, London. pp 1-19
71. Hodak C (1999) Les animaux dans la cité: pour une histoire urbaine de la nature. Genèses 37:156-169
72. Hoüard D (1780) Dictionnaire analytique, historique, étymologique, critique et interprétatif de la Coutume de Normandie. Leboucher, Rouen.
73. International Union for Conservation of Nature (2013) Invasive species. http://www.iucn.org/about/work/programmes/species/our_work/invasive_species/ Accessed 1 Oct 2013

74. Jacquin L, Cazelles B, Prévot-Julliard A-C, Leboucher G, Gasparini J (2010) Reproduction management affects breeding ecology and reproduction costs in urban feral pigeons (Columba livia). Can J Zool 88:781-787

75. Janzen DH (1985) On ecological fitting. Oikos 45:308-310.

76. Jerolmack C (2007) Animal archeology: domestic pigeons and the nature-culture dialectic. Qual Sociol Rev III(1):74-95

77. Jerolmack C (2008) How pigeons became rats: the cultural-spatial logic of problem animals. Soc Probl 55(2):72-94

78. Johnston RF, Janiga M (1995) Feral pigeons. Oxford University Press, Oxford.

79. La Fontaine J (1678) Fables, Livres VII, VIII, IX. Hatier, Paris.

80. La Perre de Roo V (1877) Le Pigeon Messager, ou Guide Pour l'Elève du Pigeon Voyageur et Son Application à l'Art Militaire. Deyrolle Fils, Paris.

81. La Ville de Mirmont H (1894) La Mythologie et les Dieux dans les Argonautiques et dans l'Enéide. Hachette, Paris.

82. Lagrue C (1897) Engrais Chimiques. Journal de Bayeux, Sommaires de leçon de Sciences, Hygiène et Agriculture du Bessin. Bayeux. pp 62-67

83. Latour B (1993) We have never been modern. Harvard University Press, Cambridge.

84. Leblanc N (2003) La place de l'animal dans les politiques urbaines. Communications 74:159-175.

85. Lépine P, Sauter V (1952) Sur l'infection des Pigeons Parisiens par le Virus de l'Ornithose. Bull Acad Natl Med 135:332

86. Lings M (1986) Le prophète Muhammad, sa Vie d'Après les Sources les Plus Anciennes. Editions du Seuil, Paris.

87. Lowe SJ, Browne M, Boudjelas S (2000) 100 of the world's worst invasive alien species. IUCN/SSC Invasive Species Specialist Group (ISSG), Auckland, New Zealand.

88. Lunel B (1858) Colombe. Dictionnaire universel des Connaissances Humaines. Lacroix-Comon, Paris, France.

89. Marion L (2013) Is the Sacred ibis a real threat to biodiversity? Long-term study of its diet in non-native areas compared to native areas. C. R. Biologies. http://dx.doi.org/10.1016/j.crvi.2013.05.001.

90. Menon (1749) La Science du Maître d'Hôtel Cuisinier, avec des Observations sur la Connaissance et Propriétés des Alimens. Paulus du Mesnil, Paris.
91. Micoud A (2010) Sauvage ou domestique, des catégories obsolètes? Sociétés 2(108):99-107
92. Miguel I, Valentim J-P, Carugati F (2010) Intelligence and its development: social representations and social identities. Pap Social Representations 19:20.1-20.33
93. Minteer BA, Collins JP (2005) Why we need an 'ecological ethics'. Front Ecol Environ 3(6):332-337
94. Moloney G (2010) Articulating Social Representations and Identity through Process & Content. Papers on Social Representations 19:15.1-15.16
95. Mooney HA, Mack RN, McNelly JA, Neville LE, Schei PJ, Waage JK (2005) Invasive alien species. Island Press, Washington, A new synthesis.
96. Moquin-Tandon A (1860) Éléments de Zoologie Médicale, Contenant la Description des Animaux Utiles à la Médecine et des Espèces Nuisibles à l›Homme. Baillière, Paris.
97. Moscovici S, Hewstone M (1983) Social representations and social explanations: from the 'Naïve' to the 'amateur' scientist. In: Hewstone M (ed) Attribution theory: social and functional extensions, Blackwell, Oxford. pp 98-125
98. Mullin MH (1999) Mirrors and windows: sociocultural studies of human-animal relationships. Annu Rev Anthropol 28:201-224
99. Nunez MA, Bailey JK, Schweitzer JA (2010) Population, community and ecosystem effect of exotic herbivores: a growing global concern. Biol Inv 12:297-301
100. O'Rourke E (2000) The reintroduction and reinterpretation of the wild. J Agric Environ Ethic 13:145-165
101. Orton D (2010) Both subject and object: herding, inalienability and sentient property in prehistory. World Archaeology 42(2):188-200
102. Palliez L (1932) Le Pigeon Voyageur. Son Origine, Description, ses Qualités, ses Aptitudes. Son Utilisation au Cours des Guerres Anciennes et Moderne. G. Baratte, Lille.

103. Philo C, Wilbert C (eds) (2000) Animal spaces, beastly places Routledge, New York.
104. Pierre E (2007) Réformer les relations entre les hommes et les animaux: fonction et usages de la loi Grammont en France (1850–1914). Déviance et Société 1(31):65-76
105. Pimentel D, McNair S, Janecka J, Whitman J, Simmonds C, O'Connell C, Wong E, Russel L, Zern J, Aquino T, Tsomondo T (2001) Economic and environmental threats of alien plant, animal, and microbe invasions. Agric Ecosyst Env 84:1-20
106. Prevost P (2006) Les Bases de l'Agriculture. Lavoisier, Paris.
107. Primack RB, Sarrazin F, Lecomte J (2012) Biologie de la conservation. Dunod, Paris.
108. Renauldon J (1765) Traité historique et pratique des droits seigneuriaux. Knapen, Paris, France.
109. Rengade J (1887) Les Besoins de la Vie et les Eléments du Bien-être, Traité Pratique de la Vie Matérielle et Morale de l›Homme. La Librairie Illustrée, Paris.
110. Reutter L (1916) Des remèdes d'Origine Humaine et Animale Prescrits au Temps des Romains en Europe (suite et fin). Bulletin de la Société d'Histoire de la Pharmacie 13:201-204
111. Ribémont B (1992) Les Oiseaux Domestiques dans les encyclopédies médiévales: quelques exemples 'à la ferme'. Proceedings of the Symposium L'homme, l'animal domestique et l'environnement, Nantes 22–24 October 1992, R. Durand. 291-304
112. Richelet (1680) Colombe. Dictionnaires des 16è et 17è Siècles. Huguet, Paris.
113. Robbins P (2007) Lawn people. How grasses, weeds, and chemicals made us who we are. Temple University Press, Philadelphia.
114. Royer B (1983) Elevage du Pigeon de Chair. du Point Vétérinaire, Maisons-Alfort.
115. Rupin E (1884) Droit de Colombier. Bulletin de la Société scientifique historique et archéologique de la Corrèze. Société Scientifique, Historique et Archéologique de la Corrèze. 314-321
116. Russell N (2002) The wild side of animal domestication. Soc Anim 10(3):285-302

117. Saint-Hilaire G (1836) Acclimatation et Domestication des Animaux Utiles. La Maison Rustique, Paris.
118. Salo P, Korpimäki E, Banks PB, Nordström M, Dickman CR (2006) Alien predators are more dangerous than native predators to prey populations. Proc Roy Soc B 274:1237-1243
119. Seastedt T, Hobbs RJ, Suding KN (2008) Management of novel ecosystems: are novel approaches required? Front Ecol Environ 6:547-553
120. Smith N (2006) Foreword. In: Heynen N, Kaika M, Swyngedouw E (eds) The Nature of cities: urban political ecology and the politics of urban metabolism, Routledge, London.
121. Sol D (2008) Artificial selection, naturalization, and fitness: Darwin's pigeons revisited. Biol J Linn Soc 93:657-665
122. Spalikowski E (1895) Les Superstitions Médicales Normandes. Bulletins de la Société d'Anthropologie de Paris IV(6):476-478
123. Stuart SN, Chanson JS, Cox NA, Young BE, Rodrigues ASL, Fischman DL, Waller RW (2004) Status and trends of amphibian declines and extinctions worldwide. Science 306:1783-1786.
124. Swyngedouw E (1999) Modernity and hybridity: nature, regeneracionismo, and the production of the Spanish Waterscape, 1890–1930. Ann Assoc Am Geogr 89(3):443-465
125. Swyngedowu E (2004) Social power and the urbanization of water. Oxford University Press, Oxford.
126. Thiébaut de Berneaud A (1841) Nouveau manuel complet du cultivateur français, ou l'Art de bien cultiver les terres, de soigner les bestiaux. Roret Editeur, Paris.
127. Thomas K (1984) Man and the Natural World. Penguin Books, Harmondsworth.
128. Vacher M (1970) Les Pigeons de Paris. Réductions du Nombre de Pigeons dans Paris, Ses Problèmes, Essais d'un produit Chimique Inhibant la Ponte chez le Pigeon. Thesis Ecole Nationale Vétérinaire Alfort, Paris.
129. Valéry L, Fritz C, Lefeuvre JC, Simberloff D (2008) In search of a real definition of the biological invasion phenomenon itself. Biol Inv 10:1345-1351
130. Valéry L, Fritz H, Lefeuvre JC, Simberloff D (2009) Invasive species can also be native. Trends Ecol Evol 24:585

An alternative view is that pastoralism in Africa (and much of Central Asia) has been compatible with biodiversity for hundreds and even thousands of years and does not necessarily cause degradation (e.g., Homewood and Rodgers 1984). Currently, there is a growing consensus that pastoralist livestock husbandry practices (at least at moderate intensity) are less damaging to rangeland resources than previously thought (Boyd et al. 1999; Reid 2012). Indeed, recent research has suggested that moderate livestock densities may even be beneficial to biodiversity (Gregory and Sensenig 2010; Soderstrom and Reid 2010; Augustine et al. 2011; Woodroffe 2011;Reid 2012). Here, we briefly review these alternate views and then summarize the scientific evidence emerging from projects associated with the Kenya Long-term Exclosure Experiment (KLEE) that shed light on the relationship between livestock (and various practices associated with livestock husbandry) and biodiversity (particularly wild ungulates) in the savannas of eastern and southern Africa.

Pastoralism and Biodiversity

It is widely believed that rising numbers of pastoralist-owned livestock have led to rangeland degradation (Herlocker 1996) and a loss of biodiversity (Prins 1992,2000; Reid 2012). In particular, livestock are blamed for the documented declines in large mammalian wildlife species in Africa (Ottichilo et al. 2000; Georgiadis et al. 2007; Ogutu et al. 2009,2011). Pastoral practices in Africa have also been associated with increased bush encroachment (Archer 2010) and invasive weeds (Witt and Nongogo 2011).

At the same time, there are several pathways by which native biodiversity negatively impacts the enterprise of livestock production (Dunham et al. 2003; Sitters et al. 2009). First, wildlife competes with livestock, consuming forage resources, altering livestock behaviors, and reducing livestock productivity (Odadi et al. 2011). Second, pastoralists lose livestock—ranging from sheep and goats to cattle and camels—to wild predators (Atickem et al. 2010; Zimmerman et al. 2010). Lastly, there are complex disease interactions among wildlife, livestock, and other domestic animals, with negative repercussions for all of these guilds (see Grootenhuis 1999; Osofsky et al. 2005; Kock et al. 2010).

Despite these conflicts between biodiversity and livestock, the fact that African savannas have until recently maintained both thriving pastoral economies and densities of large mammalian wildlife (including both herbivores and predators) greater than anywhere else on earth is testament to their potential compatibility (du Toit et al. 2010; Reid 2012). Moreover, much of the argument that livestock and wildlife are incompatible is based on inference and anecdote, rather than experimental evidence. Our research, based on several replicated, large-scale experiments, suggests various positive and negative relationships between livestock husbandry and biodiversity. Specifically, our experimental work explores four broad topics:

- Livestock and wild ungulate herbivores share a common resource base, but there have been no experiments detailing how this plays out for both herbivore guilds in terms of direct and indirect competition and possible facilitation (positive interactions).
- Livestock in large parts of eastern and southern Africa are often housed in temporary corrals (bomas) each night that, once abandoned, develop into long-term ecosystem "hot spots", with a variety of possible consequences for wild herbivores and other taxa.
- Extensive burning of rangelands has until recently been commonly practiced by pastoralists (and commercial ranchers). Burned areas have the potential to provide important grazing habitats for many species of wild herbivores.
- • Boma construction, burning, and other human activities such as fuelwood harvesting, reduce tree densities. Conversely, woody plant cover is increasing in many rangelands, a change that has been attributed both to heavy grazing and increasing atmospheric CO_2. Variation in tree density may have profound implications for wild herbivores in these mixed-use landscapes.

Through reviewing our findings about these interactions among wild herbivores, livestock, and various practices commonly employed by pastoralists, our purpose is to illustrate the diversity of positive and negative interactions and suggest how pastoralism and biodiversity conservation may be compatible.

Caveat

Pastoralism as a livelihood in Africa can take a number of forms, ranging from a truly nomadic to a mostly sedentary lifestyle, and from low to high stocking densities. Pastoralists keep a variety of different livestock, including cattle, sheep, goats, camels, and donkeys. Thus, "pastoral management" can mean many different things. Moreover, most pastoralist systems are dynamic, responding to changing conditions brought about by bio-physical factors such as rainfall, as well as socio-economic factors such as access to markets, security issues, or changes in land tenure or lifestyle often imposed by outsiders. Our intent here is not to discuss "pastoralism" as a fixed management system. Further, we acknowledge that no controlled experiments can mimic land management as practiced by diverse people across large areas of land. Rather, we illustrate some of the possible impacts that may be brought about by particular activities in which many pastoralists in eastern Africa (and to some extent southern Africa) engage. These activities include livestock grazing (here, we focus on cattle grazing), constructing corrals where livestock spend the night, using fire to improve or manage forage for livestock, and cutting wood for domestic use or commercial charcoal production.

Laikipia and KLEE

We have been studying the interactions among livestock, wild ungulate herbivores, and the land they share at the Mpala Research Centre and Ranch in Laikipia, Kenya, since 1992 (Figure 1). The landscape in this region is a semi-arid savanna dominated by perennial grasses and *Acacia* trees. This region is rich in wildlife and also supports numerous private cattle ranches as well as Maasai and Samburu pastoralists living on "group ranches". Common wild ungulate species in this region include elephant, plains zebra, Grant's gazelle, impala, eland, oryx, hartebeest, and buffalo. Common large predators include lion, leopard, cheetah, and hyena. The region is also home to critical populations of endangered African wild dog, Grevy's zebra, and black rhinoceros. Unlike other parts of Kenya, this region has seen an increase in the populations of many wildlife species in recent decades (Georgiadis 2011; M. Kinnaird et al., unpublished data). Although these increases have largely occurred on privately owned ranches, innovative conservation and tourism partnerships between these ranches and pastoralist communities raise hope that these trends will continue into the future (Sundaresan and Riginos 2010).

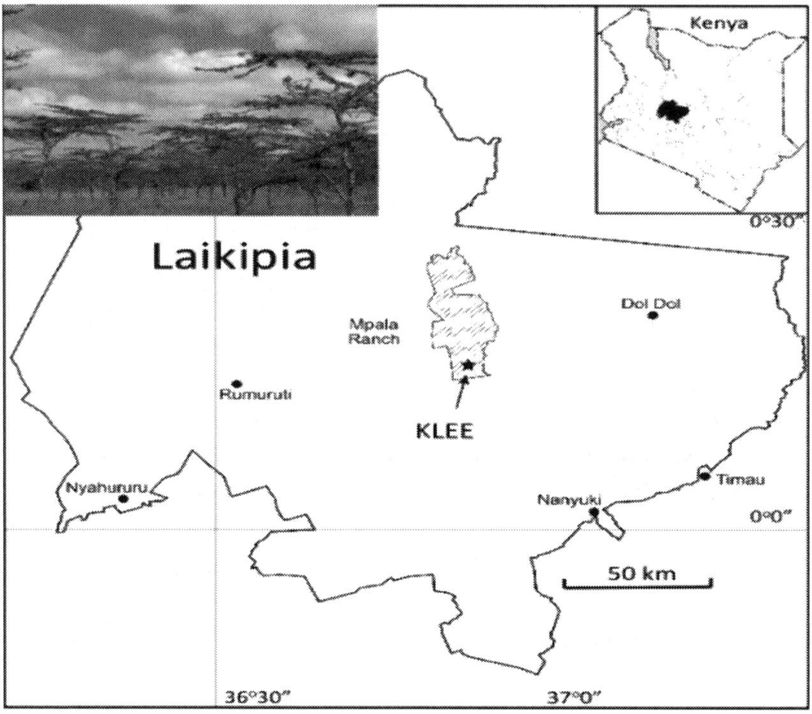

Figure 1: Laikipia County (District) in Kenya, with the location of the Mpala Ranch and Conservancy, and the KLEE exclosures. Inset:*Acacia drepanolobium*-wooded grassland.

The focus of this study in this system is the Kenya Long-term Exclosure Experiment (KLEE). This large-scale, experiment consists of eighteen 4-ha plots that were established in 1995. The plots allow six combinations of (1) cattle, (2) wild herbivores >15 kg, and (3) megaherbivores (elephants and giraffes), replicated across three blocks. These three herbivore classes are labeled C, W, and M, respectively, in Figure 2 (with "O" representing no large herbivores). Wild herbivores are allowed to come and go freely (except where they are experimentally excluded). Dung surveys suggest that wild ungulates use areas inside and outside of the plots at similar rates. Cattle use of the plots is regulated through regular "cattle runs", where herds of approximately 120 cattle are individually herded into designated plots approximately 6 times per year, which approximates the (moderate) stocking densities and return intervals of cattle on the ranch (0.15–0.20 cattle/km^2). For details of the experimental design, see x(du Toit et al. 1998).

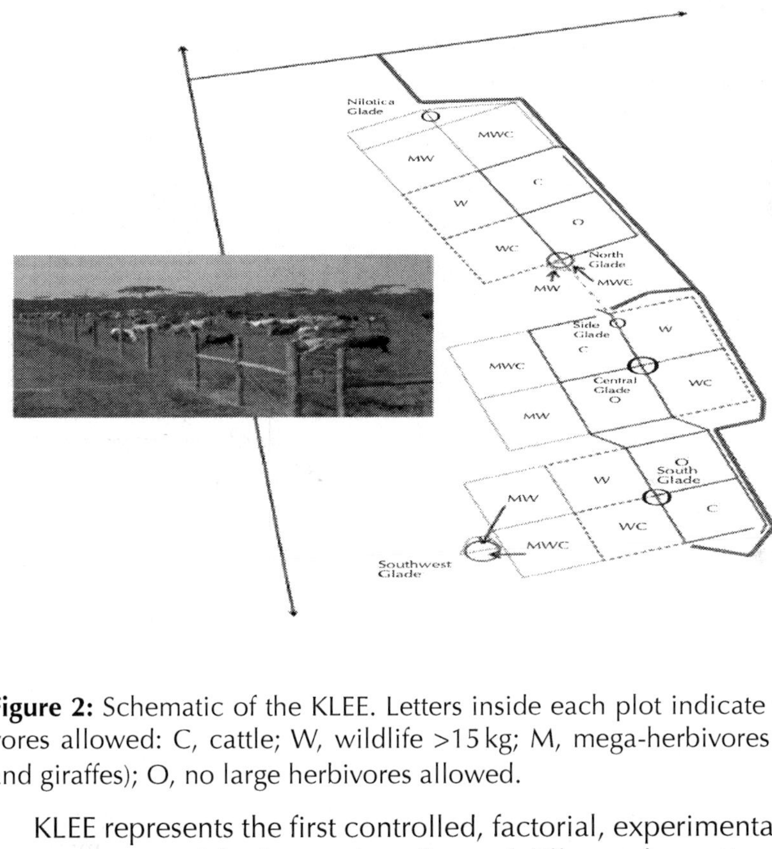

Figure 2: Schematic of the KLEE. Letters inside each plot indicate the herbivores allowed: C, cattle; W, wildlife >15 kg; M, mega-herbivores (elephant and giraffes); O, no large herbivores allowed.

KLEE represents the first controlled, factorial, experimental study in any ecosystem of the interactive effects of different domestic and native herbivores on each other and on their shared ecosystem. This study has also been extended to include several other experiments in the same landscape that manipulate fire, bomas, and tree density. Descriptions of these experiments are given below.

KEY FINDINGS

Impacts of Cattle on Wildlife

In over 15 years of work in KLEE, we have found that cattle suppress a broad spectrum of wild herbivore species, presumably through competition for shared forage resources. Wild herbivore use of these plots has been measured by counting dung piles along permanent transects that have been sampled twice-yearly since 2006. Dung density of all species was, across all sample periods 2006–2011, 44% greater in plots from which cattle have been excluded. Cattle reduced herbivore use for each of the four most common, medium-sized wild herbivore species: plains zebra, Grant's gazelle, eland, and oryx (Figure 3). Although cattle are primarily grazing herbivores, their diet at our study site can be as much as 15% forbs (non-grass herbs) (Odadi et al. 2007), and this is reflected in their ability to competitively suppress not only grazing wildlife, but also mixed feeders. None of these reductions is likely due to direct interactions with cattle or herders; cattle herds are present in the designated plots for only 12–16 h per year. Instead, cattle are consuming forage that would otherwise support wild herbivore species.

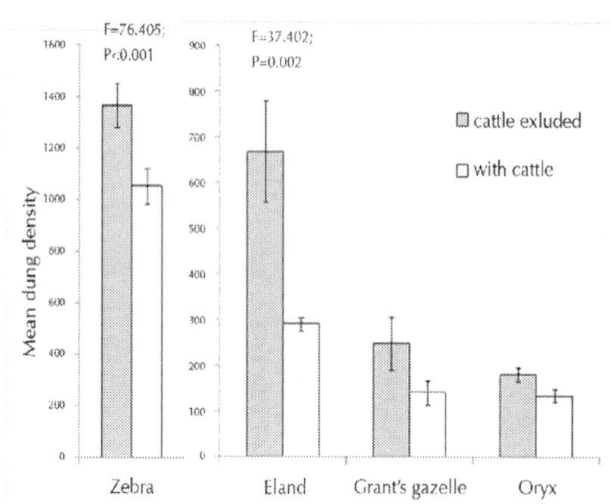

Figure 3: The abundance of herbivore dung in plots accessible to wildlife only (W and MW; shaded), and accessible to both wildlife and cattle (WC

and MWC; open), across all surveys 2006–2011.Error bars are standard errors, with a sample size of three experimental blocks.

Mega-herbivores (elephants, mainly) can potentially mitigate this competition between cattle and other wild herbivores. In the absence of elephants, cattle reduced overall wild herbivore density in 2006–2011 by 39%, whereas when elephants were present cattle reduced wild herbivore density by only 26%. In a previous study in KLEE, cattle reduced zebra use of the plots by 44% in the absence of elephants, but cattle reduced zebra use by only 18% in the presence of elephants (Young et al. 2005). In other words, elephants ameliorate the competition between cattle and wild herbivores, especially zebras. This surprising indirect facilitation is likely because (a) elephants eat forage that other wild herbivores might otherwise eat, reducing habitat use by zebras in the absence of cattle, and (b) cattle appear to eat less (reduce the forage less) when they share their forage base with elephants, increasing habitat use by zebras. Taken together, these results illustrate that grazing by domestic livestock does competitively suppress wild ungulate herbivores, but perhaps to a lesser degree than previously thought.

Special Story: Rodents

Another set of actors ameliorating competition between cattle and wild herbivores lies at the opposite end of the size spectrum: rodents. These "cryptic consumers" (Keesing 2000) also increased in plots from which we have excluded large herbivores. In plots where all large herbivores (both cattle and wild herbivores) are excluded, rodents have doubled in numbers, eating an estimated half of the forage that is made available (Keesing 1998). When only cattle or only wild herbivores are excluded, the rodent response is intermediate (Goheen et al. 2010). This means that when managers estimate how much additional forage will be available if they eliminate grazing wildlife, they are likely to over-estimate the benefits by an appreciable margin (perhaps 25%). Conversely, estimates of how much more wild herbivore biomass an ecosystem would support if all livestock were removed would be similarly over-estimated if rodents are not considered.

Impacts of Wild Ungulate Herbivores on Cattle

Another set of findings from our work in KLEE indicates that wild herbivores competitively suppress cattle, but again, to a lesser degree than scientists and managers have often assumed. In a series of experiments, we assessed the effects of wild herbivores on cattle foraging behavior, nutrition, and performance (weight gain) during dry and wet seasons (Odadi et al. 2007, 2009, 2011). We also measured vegetation cover (the forage available to the cattle) during these trials. Our results (Table 1) show that the effects of wild herbivores on cattle differ markedly between wet and dry seasons.

Table 1: Effects of the presence (versus exclusion) of wildlife on the foraging behavior, forage selection, food intake, diet quality and weight gain by cattle during wet and dry seasons

	Dry season	Wet season
Foraging behavior		
Bite rate	Decrease	No effect
Step rate	Increase	No effect
Bites per step	Decrease	No effect
Relative bites on forbs	Decrease	No effect
Relative bites on P. stramineum	Decrease	No effect
Food intake	Decrease	No data
Diet quality		
Dietary crude protein content (CP)	No effect	Increase
Dietary digestible organic matter content (DOM)	No effect	No effect
Dietary DOM/CP ratio	No effect	Decrease
Performance		

Weight gain	Decrease	Increase
Herbaceous vegetation		
Overall grass cover	Decrease	No effect
Cover of forbs	Decrease	No effect
Cover of P. stramineum	Decrease	Decrease
Cover of dead stems	No effect	Decrease

"No effect" means no statistically significant effect.

Riginos et al.

Riginos et al. Pastoralism: Research, Policy and Practice 2012 2:10 doi:10.1186/2041-7136-2-10

During the dry season, wild herbivores had a number of negative effects on cattle. Cattle bite rate decreased and step rate increased, leading to fewer bites per step. This was associated with reduced forage, reduced overall food intake, and reduced relative intake of forbs and the perennial grass *Pennisetum stramineum*. As a consequence of these changes in foraging behavior, weight gain by cattle also decreased when cattle shared forage with wild herbivores during dry periods (Figure 4). Cattle that grazed in areas not shared with wild herbivores gained weight 37% faster.

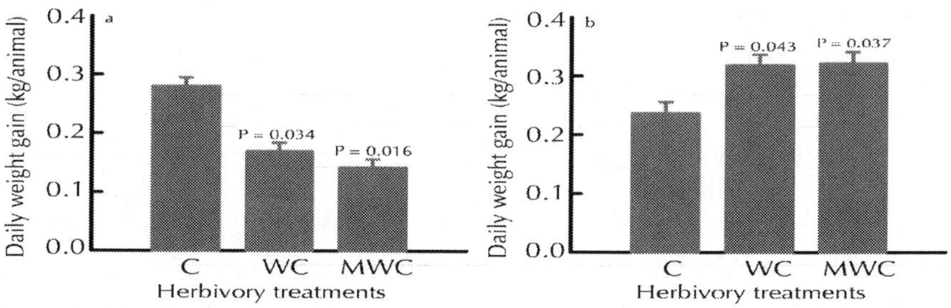

Figure 4: Cattle weight gain in the presence and absence of wildlife, during (a) dry periods and (b) wet periods. Figure adapted with permission from Odadi et al. (2011).

Decreased bite rate and bites per step and increased step rate in the shared plots suggests that cattle sharing land with wild herbivores forage less efficiently during dry periods because they travel more between feeding stations and take fewer bites at each station (see also Ungar and Noy-Meir 1988; Spalinger and Hobbs 1992; Bradbury et al. 1996). These wildlife-driven effects on foraging patterns appear to cause cattle to expend more energy and thus gain less weight (Odadi et al. 2011). Decreased consumption of forbs (non-grasses) by cattle when they forage in the same areas as wild herbivores is related to reduced availability (cover) of this forage class. Forbs contain higher protein levels than grasses (Boutton et al. 1988; Kinyamario and Macharia 1992) and are thus an important forage resource for herbivores. Although cattle are primarily grazers, inclusion of forbs in their diet may enhance their nutritional status, especially during the dry season when nutrient content of grasses is low. Depressed cattle weight gain in the shared plots during the dry season was also associated with decreased food intake and reduced relative consumption of the high protein-content grass *P. stramineum*. These findings provide some of the first experimental evidence to support the long-held assumption that wild herbivores compete with cattle for food.

However, the story is not one of simple competition. During the wet (growing) season, the presence of wild herbivores enhanced cattle weight gain (Figure 4), indicating a facilitative interaction (Odadi et al. 2011). This positive interaction was driven by improved forage quality due to reduced cover of dead grass stems in areas where wild herbivores also grazed. This facilitation appears to be largely driven by zebras, which are known to consume dead grass parts and have been observed to enhance the quality of forage available to other wild herbivores (Gwynne and Bell 1968; Duncan et al. 1990).

These findings indicate that although grazing wild herbivores do compete with cattle in dry periods, they at least partially compensate for this by facilitating cattle in wet periods. These findings suggest that wild herbivores are not uniformly detrimental to livestock production, improving the prospect that wildlife conservation and economic development can both be achieved in savanna rangelands.

Long-Term Effects of Traditional Livestock Bomas

Throughout the savannas of eastern Africa and parts of southern Africa, bomas (temporary thorn-fence livestock corrals; Figure 5a) have been used for centuries by pastoralists to corral livestock overnight for protection from both predators and stock raiders (Western and Dunne 1979). During the colonial era (a time of suppressed lion populations), traditional bomas on many of Laikipia's commercial ranches were replaced by (non-thorn) wooden fence corrals, presumably because the threat of livestock predation declined (P. Jessel, personal communication). In recent decades, predator populations (including lions) have begun to recover in some parts of Laikipia, and thorn bomas continue to be an important element of economic and cultural life on both communally and privately managed lands.

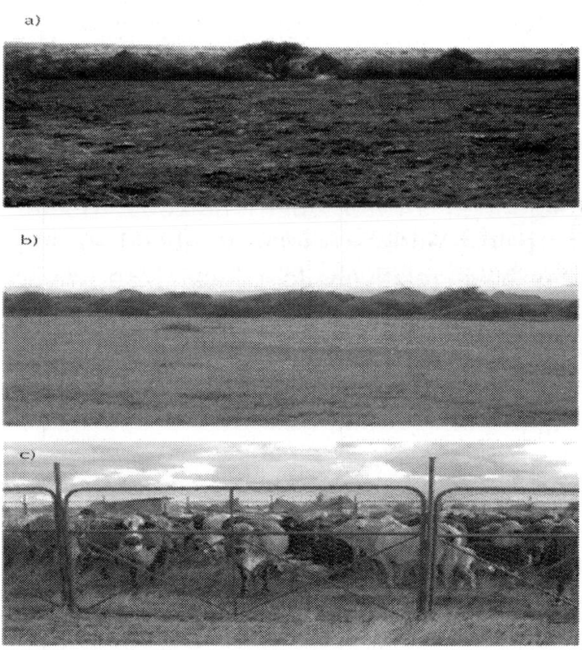

Figure 5: (a) Traditional thorn-fence bomas (livestock corrals) are used for months to years at a time. (b) Treeless glades develop on boma sites and per-

sist for several decades as hotspots of high-quality grass that attract intense wildlife use. (c) Metal "mobile bomas" are a new adaptation of thorn-fenced bomas.

A given boma may be used for months or years at a time before a new boma is established elsewhere (Western and Dunne 1979). While they are actively being used, bomas can have a number of negative impacts on wild herbivores and other taxa. Scores of trees in the vicinity are harvested to build the boma fences. The intensive grazing around bomas can cause overgrazing, soil compaction and, over time, land degradation (Ringrose et al. 1996; Moleele et al. 2002;Muchiru et al. 2009). At the same time, wild herbivores avoid active boma sites, either because of the presence of people or competition for shared resources, or both (Sundaresan et al. 2007). Depending on the duration of boma use and the severity of the livestock's impacts on surrounding land, negative effects of a boma may persist for years, possibly even decades, after it is abandoned.

However, boma practices also have a number of more complex, often positive, impacts on wild herbivores and other biodiversity. Most notably, abandoned boma sites go on to become ecosystem hotspots for a variety of flora and fauna. After abandonment, a thick layer of livestock dung is left at the site, facilitating the establishment of nutrient-rich plant communities that persist for decades to centuries (Blackmore et al. 1990; Reid and Ellis 1995; Augustine 2003; Muchiru et al. 2009) and creating a treeless landscape feature called here a "glade" (Figure 5b). Glade sites have relatively low local plant diversity but their presence increases regional diversity by hosting plant species that are uncommon in or absent from the background savanna (Muchiru et al. 2009;Porensky 2011; Veblen 2012). Glades attract wild herbivores and cattle to graze on nutrient-rich grasses and rest in these open areas (Young et al. 1995; Augustine 2004; Muchiru et al. 2008;Veblen 2012). Our work has also shown that glades host increased densities of termites (Veblen 2008a), arboreal geckos and insects (Donihue paper is in revision at Ecology), and locally rare rodent species (Veblen 2008a). All of these results suggest that bomas and glades represent a rarely documented example of a positive, synergistic relationship between biodiversity conservation and livestock production. Because glade edge effects extend at least 100-m into the surrounding savanna landscape (Young et al. 1995; Muchiru et al. 2009; Porensky 2011;

Veblen 2012), these anthropogenic features influence extensive areas across the rangelands of eastern and southern Africa.

Our work in KLEE (which encompasses several glades) indicates that cattle and wild herbivores (particularly medium-sized herbivores) have contrasting impacts on herbaceous plant community succession in glades (Veblen and Young 2010). Over the long term, wild herbivores (mostly zebras), through their forage preferences, reinforce landscape heterogeneity by helping to maintain glades in a more palatable, early successional state. Cattle and mega-herbivore wildlife, on the other hand, appear to reduce this effect by preferentially foraging on early successional grass species, thereby favoring the later successional species (Veblen and Young 2010).

We have also found that cattle grazing alters the use of glades by wild herbivores. During the dry season, wild herbivore use of glades more than doubled when cattle were excluded (Porensky et al. in review). This response was driven by both grazing herbivores (zebra, hartebeest, and warthog) and mixed-feeding or browsing herbivores (eland, oryx, steinbuck, Grant's gazelle, bush duiker, and impala). Moreover, in the presence of cattle, wild herbivores showed little preference for glades (compared to the background savanna), but when cattle were excluded, wild herbivore use inside glades was at least 30% higher than their use of areas far from glades. Other work (Veblen 2008b) suggests that glades may serve as important wild herbivore foraging areas during the dry season, when grass is scarce in non-glade areas (*sensu* Ilius and O'Connor).

Together, these results from KLEE suggest that different types of herbivores (cattle, medium-sized wild herbivores, and mega-herbivores) interact in complex ways with glades. Through the creation of glades and by facilitating subsequent interactions between cattle and glades, traditional pastoralist management activities have had major impacts on savanna landscape heterogeneity and biodiversity.

Bomas in the Modern Context

As pastoral populations have grown and become more sedentary and the land available to them shrinks, livestock management practices and the ecological impacts of bomas are changing. Most pastoralists today cannot move as far or as frequently as their predecessors (Ensminger and Rutten 1991; Fratkin et al. 1999), and herders are forced to use smaller parcels of land more intensively. As a result of these processes, boma densities have increased dramatically in some parts of East Africa (Lamprey and Reid 2004), with detrimental impacts on widlife (Ogutu et al. 2010). Moreover, without sufficient land or mobility, livestock managers have difficulty allowing for the full recovery of intensively used areas around traditional boma sites.

In this context of increasing boma density and duration of use, bomas may lead to significant degradation of surrounding soils and plant communities. We have found that areas between two or more nearby glades (<150 m apart) are characterized by especially high tree densities and low wild herbivore use (Porensky 2011). This is likely due to a legacy of active boma use; the presence of cattle and humans during boma occupation may positively affect tree cover by reducing grass cover (and its competitive effects on trees) and browser presence, and by fertilizing areas between nearby bomas (L. M. Porensky unpublished data; Muchiru et al. 2009). Following boma abandonment, high tree densities between nearby glades probably deter wild herbivores by making predator detection more difficult (see below).

We have been exploring ways to reduce negative—and amplify positive—impacts of modern boma management. One potential tool that has been developed in Laikipia is the "mobile boma". Mobile bomas are made of moveable, metal fencing (Figure 5c) and can be used for short periods of time so that they have fewer degradative impacts but still create nutrient-rich grazing hotspots. Mobile bomas can be moved more frequently than traditional brush bomas because they do not require the labor or large number of trees that are needed to construct brush bomas. More frequent herd movement reduces overgrazing and trampling at any one site, potentially allowing areas to recover and be reoccupied sooner.

The advent of mobile bomas raises several key management questions and opportunities. Managers in Laikipia are now using mobile bomas to create wildlife hotspots for tourists and to restore grass in degraded, bare areas. Questions that remain to be answered about this new management tool include: Is there a minimum time that a boma must remain in place in order to create a wildlife hotspot? How often does a boma need to be moved in order to avoid over-use and degradation of a given site? How should managers arrange bomas across the landscape to minimize broad-scale degradation? We are currently investigating these questions experimentally with a view to providing practical recommendations for how managers can adapt this traditional management practice to improve forage quantity and quality for their livestock and provide nutrient hotspots for wildlife.

Fire

Fire has been structuring savanna and grassland systems for millennia. Gillson (2004) has recorded charcoal evidence of fire history over the last 1,400 years in Kenya. In Laikipia, burning was a common practice among both pastoralists and commercial ranchers until the early 1960s (Heady 1960). In focus group interviews, Maasai, Samburu, and Borana elders in Kenya and Ethiopia recalled burning every several years to control bush encroachment, control tick and tsetse fly populations, and remove rank or overgrown grass with a high ratio of dead stems—allowing newer, more palatable grass to grow in its place (C. Riginos, unpublished data). At present, most pastoralists and commercial ranchers in the region do not use fire as a management tool. On pastoralist-managed lands, this is largely because current grass cover is insufficient fuel to carry a fire and because people are hesitant to burn their dry season reserves. Most elders we have interviewed, however, agree that controlled burns offer a number of benefits to livestock and wildlife and claim they would resume burning if grass were not in such short supply.

To investigate the impacts of fire on wildlife in this region, we used multiple replicated prescribed burns on several commercial ranches in Laikipia to create habitat heterogeneity at multiple scales (Figure 6). We then measured the use of these sites by wild herbivores using repeated dung surveys. We also measured changes in grass quality and quantity in burned sites to estimate how forage dynamics influenced wild herbivore habitat use, with particular interest in determining whether a grazer's body size influenced its time spent in burned areas.

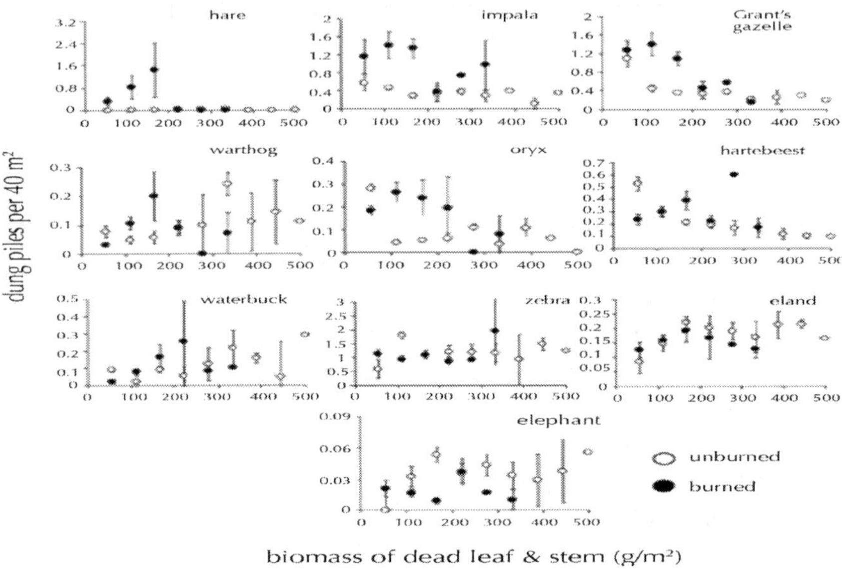

Figure 6: Hartebeest, zebras, and giraffe attracted to a 1-ha burn in Laikipia, Kenya.

We found that prescribed burning creates heterogeneity in forage quality and quantity across the landscape that persists for several years after burning (Sensenig et al. 2010). The use of burn sites by wild herbivores does depend on a grazer's body size, with small-bodied animals using burn sites significantly more than large-bodied grazers (Sensenig et al. 2010). Foregut fermenters (ruminant herbivores) prefer burns more than hindgut fermenters (non-ruminants, e.g., zebras, elephants, and hares) of similar body size (Sensenig et al. 2010). This suggests that burns are especially important for maintaining populations of small-bodied and ruminant wild herbivores.

These herbivore responses to burning are also of management importance in many parts of eastern Africa where significant wild herbivore declines have been observed (Ottichilo et al. 2000). There is growing evidence that in some cases wild herbivore declines may be due to reduced grass structural heterogeneity—due to fire suppression in an otherwise fire-prone system—which may have negative repercussions for certain grazing herbivores. For example, over the last 20 years, Thomson's gazelle, Grant's gazelle, and wildebeest have declined in the Ngorongoro Crater in Tanzania, while the bulk forage feeder, the African buffalo, has increased (Estes et al. 2006). Fire exclusion since 1960s was cited as one of the multiple potential causes, which led to an increase in tall, poor quality grass at the expense of shorter, high-quality forage. Similar scenarios have been suggested for Nairobi National Park, Kenya (Gichohi 1992).

In a separate study in our ecosystem, we also demonstrated the effectiveness of fire in reducing woody vegetation (Okello et al. 2008). Eight years after a controlled burn, tree density was reduced by 64% inside the burn site relative to adjacent unburned areas. This reduction in tree density likely further enhances the grass productivity and quality of the habitat for a variety of wild herbivore species (see below).

Although many pastoralists do not currently use prescribed burns, our work illustrates that fire may still be an important and useful management tool; particularly in regions such as Laikipia where many pastoralist communities seek to attract wildlife and tourism to their land. For example, a shifting mosaic of relatively small burns might provide many of the benefits of burning (including providing more nutritious forage for livestock and controlling bush encroachment) without incurring large losses of forage at any one time. The global trend towards increased woody cover (Wigley et al. 2010) may also provide a renewed role for fire in controlling bush encroachment and maintaining a rangeland suitable for cattle and other grazing livestock. However, it is important to note that not all areas are amenable to burns. All of the burn plots in our study were located on black cotton vertisol soils receiving 500–700 mm mean annual rainfall. Fire-induced heterogeneity of grass swards is likely more important in these kinds of habitats where significant seasonal accumulation of grass biomass and declines in forage quality occur. In areas that are dominated by bare ground and eroding rapidly, a burn might remove litter and other important ground cover, accelerating erosion and degradation. In such areas, a period of rest from grazing (allowing grass to recover) followed by a controlled burn can be very effective at controlling woody vegetation and stimulating new grass production.

Tree and Shrub Density

People and their livestock affect the cover and density of woody plants (both trees and shrubs) in a variety of ways. Pastoralists cut woody stems for a number of purposes: as fuel for domestic fires or, more recently, to make charcoal for sale in urban areas, as brush walls for bomas, for construction of houses, and for medicinal purposes. This harvesting of woody vegetation can reduce the density and cover of trees and shrubs, particularly in areas where people are partly or fully sedentary. Pastoralists' browsing livestock (e.g., goats, camels) can also have a negative impact on woody plants, ultimately reducing their numbers. Finally, as discussed above, fire can reduce woody cover and density (Okello et al. 2008).

Other activities have the opposite effect, causing woody plants to increase in number. Fire suppression and heavy grazing have both been associated with increased numbers of trees and shrubs (so-called "bush encroachment"), especially around intensely grazed areas like water holes (Moleele and Perkins 1998) and bomas (see above, and Moleele et al. 2002). Through experimental removals of grass around trees, both inside the KLEE experiment and outside, we have shown that grass can have a strong, negative (competitive) effect on trees of all sizes—halving their rate of growth (Riginos and Young 2007; Riginos 2009). At the same time, the absence of browsers (e.g., where cattle have replaced wild herbivores) also enables trees to grow and reproduce more rapidly (Riginos and Young 2007; Maclean et al. 2011; Porensky and Veblen 2012). These studies together provide new insights into how livestock grazing affects woody cover. It is also important to note that local anthropogenic activities are likely acting in concert with a global increase in carbon dioxide to favor woody plants over grasses (Wigley et al. 2010; Archer 2010).

Changing numbers of trees and shrubs can have a variety of consequences for native flora and fauna. Trees and shrubs provide essential habitat structure in the savanna landscape—for example, they provide perches and nest sites for birds (Ogada et al. 2008), food and domicile for insects, spiders, and invertebrates (Pringle et al. 2007), and a sheltered, nutrient-enriched habitat for a variety of understory plants. In the KLEE system, we have demonstrated that lower densities of a common shrub, *Cadaba farinosa*, are associated with lower abundances in several common butterfly species in the genus *Colotis*, whose adults and larvae specialize on Capparaceae (M. L. Wilkerson and T. P. Young unpublished data). Thus, an increase in trees and shrubs may be beneficial for overall ecosystem species diversity—up to a point. A very high density of woody vegetation, conversely, is likely to hinder the persistence of species that rely on open, grassy areas.

Similarly, changes in woody vegetation can have a variety of consequences for large herbivores, both wild and domestic. Increased woody vegetation may favor browsing herbivores; however, in many cases it is the less palatable trees and shrubs that tend to increase most aggressively. Through a large-scale manipulation of tree density in our study system, we have demonstrated that grass production is suppressed in areas with more trees (Riginos et al. 2009), reducing the forage available to grazing wildlife and livestock. Dense thickets of woody vegetation can also be physically impenetrable for livestock and wildlife alike. In some areas, these thickets tend to harbor tsetse flies and the diseases they carry.

Finally, our work has shown that many species of wild herbivores avoid areas of dense woody vegetation (Figure 7; Riginos and Grace 2008), most likely because the vegetation impedes their ability to detect and flee from predators. On the other hand, areas with few or no trees may not be suitable habitat for browsing herbivores or mixed-feeders. Trees, particularly larger ones, can enhance the nutrient content of the soil and grasses growing beneath them; thus, even grazing herbivores may suffer in areas from which all trees have been removed (Treydte et al. 2010).

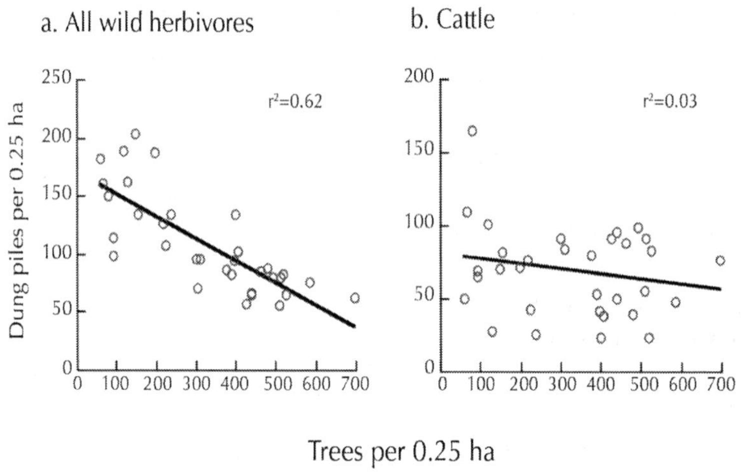

Figure 7: Dung densities indicate that wild herbivore species (predominantly plains zebra, Grant's gazelle, hartebeest, and oryx) generally prefer more open areas, while cattle use the landscape more indiscriminately. Figure printed with permission from Riginos and Grace (2008).

Cattle (which are typically accompanied by herders in eastern Africa) appear not to be as sensitive to tree density as are wild herbivores (Figure 7; Riginos and Grace 2008; W. O. Odadi and C. Riginos, unpublished data), suggesting that competition between these guilds of herbivores may be lessened by focusing livestock grazing on more densely wooded areas. During times of drought, however, many wild herbivore species that prefer open areas shift into more wooded areas if there is more grass in those areas (C Riginos, unpublished data). These results again highlight that season and habitat are important determinants of the strength of competition between wild herbivores and livestock.

The many ways in which pastoralists impact woody vegetation—and therefore wild herbivores and other biodiversity—suggest both an opportunity and a caution. Because people and their livestock can have numerous, opposing effects on woody cover, there is good reason to believe that the needs of both pastoralists and wild herbivores can be met in their shared habitat. A moderate density of a variety of tree and shrub species is most likely to benefit people, livestock, and a diverse community of wildlife. However, some heterogeneity in woody cover—including some densely wooded areas as well as open, grassy areas in addition to moderately woody areas—is most likely to benefit the greatest diversity of wildlife and other flora and fauna.

DISCUSSION

Our results confirm that the relationship between livestock husbandry and wildlife is complex. On the one hand, we provide some of the first evidence from replicated, controlled experiments that cattle do reduce habitat use by native herbivores (even at moderate cattle stocking densities), and that wild herbivores can reduce livestock productivity. This competition extends to browsing and mixed-feeding herbivores, suggesting that estimates of competition based on dietary overlap may underestimate its strength; even partial overlap on minority components of the diet may be important (limiting). On the other hand, our research has begun to reveal a wide variety of more positive relationships between livestock and wild herbivores. This includes not only direct facilitation of wild herbivores by cattle (particularly during the wet season), but also several indirect benefits brought about by livestock management activities, particularly the use of bomas and fire as well as various management impacts on woody cover.

More generally, our findings suggest that pastoralists (as well as other land or livestock managers) can minimize negative interactions and maximize positive interactions through two means: first, by minimizing competition between wild herbivores and livestock during the dry season; and second, by maintaining or increasing habitat heterogeneity. Minimizing competition might be achieved through avoiding critical wildlife habitat during dry times. A more heterogeneous landscape may also afford more refugia for wildlife in times of drought. Realistically, however, it may be difficult to minimize competition and land degradation unless pastoralists destock during or in anticipation of dry periods and then later restock. Drought index insurance programs (Hazell and Hess 2010) and slaughter off-take programs (Zwaagstra et al. 2010) may provide incentives to destock where conventional, commercial market sales are not a feasible option.

Habitat heterogeneity can be maintained or increased through the mechanisms we have discussed above—fire, bomas, and altering tree cover—but only if the location, and ideally size and spacing, of these impacts varies in space and time. For example, as our research has illustrated, the increasing use of semi-permanent boma sites can be a cause of landscape degradation, whereas a more traditional shifting pattern of boma locations, durations, and spacing can minimize negative impacts on the land while increasing habitat heterogeneity and creating nutrient-rich hotspots that benefit both wildlife and livestock.

It is important to keep in mind that our results have emerged from a moderate to high productivity landscape where livestock (mostly cattle) are kept at moderate stocking rates. Our conclusions about the direct competitive and facilitative interactions between cattle and wild herbivores may take a different form in other contexts. For example, we would expect competition between cattle and wild herbivores to be more intense (and facilitation weaker) in a less productive system, or one with higher stocking densities. Competitive and facilitative dynamics may also take a different shape in systems where sheep, goats, camels, or other livestock species dominate. However, our findings on how bomas, changes in woody cover, and to some extent fire (depending on the productivity of the system and its resilience) can benefit wild herbivores would apply across a wide variety of landscapes and livestock systems.

From our interactions with pastoralists in Kenya and Ethiopia, it is clear that most are aware of the various positive and negative ways in which their activities can affect wildlife. However, few appear to be using these means to intentionally encourage wildlife, ecosystem health, or biodiversity. This is in part because the incentives to do so are absent or poorly developed, although this may be changing as outsiders are coming to value the services that rangelands can provide (see below). However, a more fundamental challenge for the future coexistence of wildlife and livestock is how to ensure moderate levels of grazing given a variety of constraints on livestock management. "Moderate grazing" typically means that 35–50% of the palatable forage is consumed before the area is allowed to recover (Holechek et al. 1999). In the past, smaller populations of mobile livestock ensured a more moderate grazing regime with adequate time for plants to recover. Today, pastoralists are more constrained in their movements and management options due to higher human and livestock populations and a variety of other factors beyond their control (e.g., land subdivision, colonial, and post-colonial land settlement schemes). These constraints will be coupled with continued increases in human population, improvements in veterinary practices, and growing expectations for large increases in the standard of living—posing a multitude of challenges for future livestock–wildlife coexistence. Even today, examples of pastoralists coexisting with wildlife (i.e., situations in which wildlife populations are not declining) are few and far between. Even the highly productive Mara/Serengeti and Ngorongoro ecosystems are experiencing large wildlife declines (Ogutu et al. 2011), probably because human and livestock densities are higher than they were historically. Wildlife increases in the Laikipia ecosystem are, at present, largely occurring on properties where livestock densities are more moderate (M. Kinnaird et al., unpublished data).

More optimistically, however, a number of initiatives and possible future scenarios give hope that pastoralists and wildlife can continue to co-exist in Africa as they have for centuries. Wildlife-based tourism enterprises (at present supported by external sources) on pastoralist-owned or -managed lands are operating with some success (both in terms of income generation and wildlife conservation) in various parts of East Africa, including Laikipia and the Maasai Mara. For example, in Laikipia and adjacent Samburu District, pastoralists on a number of group ranches have set aside conservation areas within their ranches. These areas are not grazed by domestic livestock year-round but rather are left for wildlife and as a reserve for livestock in the event of a drought. In the future, there may be formal markets for other types of ecosystem services—such as carbon sequestration or even provision of clean air and water—in pastoralist lands (World Initiative for Sustainable Pastoralism 2008).

Changing management systems and lifestyles may also portend new hope for wildlife and pastoralists to co-exist. Some previously mobile pastoralists who were encouraged to become sedentary and take ownership of small parcels of land by governments (Mwangi and Ostrom 2009) are now aggregating their holdings and returning to patterns of seasonal migration more similar to those of the past (Mwangi 2007). Particularly in arid and semi-arid ecosystems, pastoral mobility affords a more resilient and potentially more productive livestock system and is less likely to cause land degradation than a more sedentary, year-round grazing regime (Western et al. 2009a,b; Hesse and Cavanna 2010). In southern Kenya, wildlife populations fared far better where pastoral mobility was maintained compared to adjacent areas that were subdivided and settled (BurnSilver et al. 2008; Western et al. 2009a, b), and sedentarization had mixed effects on livestock production, being particularly detrimental on low-productivity sites (Boone et al. 2005;BurnSilver et al. 2008). Although the KLEE experiment was not designed to ask questions about livestock mobility, it does provide one example of wildlife coexisting with cattle in an area where cattle are grazed episodically rather than continuously. Mobility may also help to create or maintain greater landscape heterogeneity and reduce dry-season competition between wild herbivores and livestock. The possibility that traditional management practices such as seasonal migration and deliberate burns may in time be reinstated raises many possibilities for a sustainable relationship among people, livestock, and wildlife.

Additionally or alternatively, human populations in rural rangelands may decrease in years to come as people choose to move to urban centers (Young 2006) and a greater proportion of family income is generated from non-farm activities (Reid 2012). This may ease the livestock grazing pressure on the land, with benefits for wildlife. However, the demographic and economic shift away from pastoralism might not reduce overall livestock pressure on the land if, for example, urban residents (particularly the more well-to-do) continue to keep herds of livestock at their rural homes, as is the case in many parts of eastern and southern Africa. Conversely, supplemental income from wildlife may encourage greater tolerance for biodiversity and, along with more equitable sharing of benefits from livestock and greater empowerment of local communities in the national and international arena, could also help foster continued co-existence between wildlife and pastoralists (Reid 2012).

Yet another possibility is that the long-term (>50 years) future of pastoralism in Africa will be so radically different from historical and contemporary pastoralism that many projections will be moot. For example, in Botswana and Kenya, rangeland landowners are increasingly absentee and hire third party pastoralists to manage their livestock—raising a whole new set of management issues. Nevertheless, livestock production in African's arid and semi-arid ecosystems—whatever form it takes—is likely to continue for the foreseeable future. Finding a path that reconciles this production with other societal values (biodiversity, ecosystem services) will remain an ongoing challenge. Current research is revealing that this challenge need not be insurmountable.

AUTHORS' CONTRIBUTIONS

This review summarizes the work of multiple projects, with each section co-written by the people who conceived, designed, carried out, analyzed, and wrote up the original research. DK and TPY, wildlife section; FK andTPY, rodent section; WOO and TPY, cattle section; LMP and KEV, boma section; RLS, fire section; CR and MLW, tree density section. TPY wrote the first draft. CR did a major rewrite and handled all subsequent drafts. All the authors provided edits and read and approved the final manuscript.

ACKNOWLEDGMENTS

We would like to thank Frederick Erii, John Lochikuya, Matthew Namoni, Jackson Ekadeli, and Patrick Etelej for their invaluable assistance in the field. We also thank Margaret Kinnaird, Michael Littlewood, and the Mpala Research Centre and its staff for their logistical support. The KLEE exclosure plots were built and maintained by grants from the James Smithson Fund of the Smithsonian Institution (to A.P. Smith), The National Geographic Society (4691–91), the African Elephant Program of the U.S. Fish and Wildlife Service (98210-0-G563) (to T.P. Young), and the National Science Foundation (LTREB BSR-97-07477, 03–16402, and 08–16453) (to T.P. Young, C. Riginos, K.E. Veblen, and K. Caylor). The boma, fire, and tree density research were also supported by the NSF LTREB grants, as well at NSF Graduate Research Fellowships (to R.L. Sensenig, K.E. Veblen, C. Riginos, L.M. Porensky, and M.L. Wilkerson) and NSF Doctoral Dissertation Improvement Grants (to R.L. Sensenig, C. Riginos, and L.M. Porensky). Rodent research was funded bY NSF (CAREER) and the National Geographic Society (to F. Keesing). We also thank Cary Farley and two anonymous reviewers for their helpful comments on the manuscript.

REFERENCES

1. Archer, S. 2010. Rangeland conservation and shrub encroachment: new perspectives on an old problem. In Wild rangelands: Conserving wildlife while maintaining livestock

in semi-arid ecosystems, ed. Toit J, Kock R, Deutsch J, 53–97. Wiley-Blackwell, Oxford, UK.

2. Atickem, A, S Williams, A Bekele, and S Thirgood. 2010. Livestock predation in the Bale Mountains, Ethiopia. African Journal of Ecology 48: 1076–1082.
3. Augustine, DJ, TP Young, KE Veblen, C Riginos, and JR Goheen. 2011. Pathways for positive cattle-wildlife interactions in semi-arid rangelands. Smithsonian Contributions to Zoology 632: 55–71.
4. Augustine, DJ. 2003. Long-term, livestock-mediated redistribution of nitrogen and phosphorus in an East African savanna. Journal of Applied Ecology 40: 137–149.
5. Augustine, DJ. 2004. Influence of cattle management on habitat selection by impala on central Kenyan rangeland. Journal of Wildlife Management 68: 916–923.
6. Blackmore, AC, MT Mentis, and RJ Scholes. 1990. The origin and extent of nutrient-enriched patches within a nutrient-poor savanna in South Africa. Journal of Biogeography 17: 463–470.
7. Boone, RB, SB BurnSilver, PK Thornton, JS Worden, and KA Galvin. 2005. Quantifying declines in livestock due to land subdivision. Rangeland Ecology and Management 58: 523–532.
8. Boutton, TW, LL Tieszen, and SK Imbamba. 1988. Seasonal changes in the nutrient content of East African grassland vegetation. African Journal of Ecology 26: 103–115.
9. Boyd, C, R Blench, D Bourn, L Drake, and P Stevenson. 1999. Reconciling interests among wildlife and people in eastern Africa: A sustainable livelihoods approach. ODI, London.
10. Bradbury, JW, SL Vehrencamp, KE Clifton, and LM Clifton. 1996. The relationship between bite rate and local forage abundance in wild Thomson's gazelles. Ecology 77: 2237–2255. Publisher Full Text
11. Porensky, LM, SF Bucher, KE Veblen, AC Treydte, and TP Young. Large herbivores alter edge effects around ecosystem hotspots in an African savanna.
12. BurnSilver, SB, J Worden, and RB Boone. 2008. Processes of fragmentation in the Amboseli ecosystem, southern Kajiado District, Kenya. In Fragmentation in semi-arid and arid landscapes:

Consequences for human and natural systems, ed. Galvin KA, Reid RS, Behnke RH, Hobbs NT, 225–253. Springer, Berlin.

13. Dunham, KM, EF Robertson, and CM Swanepoel. 2003. Population decline of tsessebe antelope (Damaliscus lunatus lunatus) on a mixed cattle and wildlife ranch in Zimbabwe. Biological Conservation 113: 111–124.
14. Duncan, P, TJ Foose, I Gordon, CG Gakahu, and M Lloyd. 1990. Comparative nutrient extraction from forages by grazing bovids and equids: A test of the nutritional model of equid/bovid competition and coexistence. Oecologia 84: 411–418.
15. 2010. Wild rangelands: Conserving wildlife while maintaining livestock in semi-arid ecosystems,Toit J, Kock J, Deutsch J Wiley-Blackwell, Oxford, UK.
16. Ensminger, J, and A Rutten. 1991. The political-economy of changing property-rights—dismantling a pastoral commons. American Ethnologist 18: 683–699.
17. Estes, RD, JL Atwood, and AB Estes. 2006. Downward trends in Ngorongoro Crater ungulate populations 1986–2005: Conservation concerns and the need for ecological research. Biological Conservation 131: 106–120.
18. Fratkin, EM, EA Roth, and MA Nathan. 1999. When nomads settle: The effects of commoditization, nutritional change, and formal education on Ariaal and Rendille pastoralists. Current Anthropology 40: 729–735.
19. Georgiadis, NJ, JGN Olwero, G Ojwang, and SS Romanach. 2007. Savanna herbivore dynamics in a livestock-dominated landscape: I. Dependence on land use rainfall density and time. Biological Conservation 137: 461–472.
20. Georgiadis, NJ. 2011. Introduction: Conserving wildlife in Kenya's Ewaso landscape. Smithsonian Contributions to Zoology 632: 1–10.
21. Gichohi, HW. 1992. The effects of fire and grazing on grasslands of Nairobi National Park. University of Nairobi, MSc Dissertation.
22. Gillson, L. 2004. Testing non-equilibrium theories in savannas: 1400 years of vegetation change in Tsavo National Park, Kenya. Ecological Complexity 1: 281–298.

23. Goheen, JR, TM Palmer, F Keesing, C Riginos, and TP Young. 2010. Large herbivores facilitate savanna tree establishment via diverse and indirect pathways. Journal of Animal Ecology 79: 372–382.
24. Gregory, NC, and RL Sensenig. 2010. Effects of controlled fire and livestock grazing on bird communities in East African savannas. Conservation Biology 24: 1606–1616.
25. 1999. 25 Years of wildlife disease research in Kenya, Nairobi,Grootenhuis JG Kenya Agricultural Research Institute, Nairobi, Kenya.
26. Gwynne, MD, and RHV Bell. 1968. Selection of vegetation components by grazing ungulates in the Serengeti National Park. Nature 220: 390–393.
27. Hazell, PBR, and U Hess. 2010. Drought insurance for agricultural development and food security in dryland areas. Food Security 2: 395–405.
28. Heady, HF. 1960. Range management in East Africa. Kenya Department of Agriculture. Government Printer, Nairobi, Kenya.
29. Herlocker, DJ. 1996. Rangeland degradation in arid and semi-arid Africa. In East African ecosystems and their conservation, ed. McClanahan TR, Young TP, 266–267. Oxford University Press, New York.
30. Hesse, C, and S Cavanna. 2010. Modern and mobile: The future of livestock production in Africa's drylands. International Institute for Environment and Development and SOS Sahel, UK.
31. Holechek, JL, H Gomez, F Molinar, and D Galt. 1999. Grazing studies: What we've learned. Rangelands 21: 12–16.
32. Homewood, KM, and WA Rodgers. 1984. Pastoralism and conservation. Human Ecology 12: 431–441.
33. Keesing, FL. 1998. Impacts of ungulates on the demography and diversity of small mammals in central Kenya. Oecologia 116: 381–389.
34. Keesing, F. 2000. Cryptic consumers and the ecology of an African savanna. Bioscience 50: 205–216.
35. Kinyamario, JI, and JNM Macharia. 1992. Above-ground standing crop, protein content and dry matter digestibility of a tropical

grassland range in the Nairobi National Park, Kenya. African Journal of Ecology 30: 33–41.

36. Kock, R, M Kock, S Cleaveland, and G Thomson. 2010. Health and disease in wild rangelands. In Wild rangelands: Conserving wildlife while maintaining livestock in semi-arid ecosystems, ed. Toit J, Kock R, Deutsch J, 98–128. Wiley-Blackwell, Oxford, UK.
37. Lamprey, R, and RS Reid. 2004. Expansion of human settlement in Kenya's Maasai Mara: What future for pastoralism and wildlife? Journal of Biogeography 31: 997–1032.
38. Maclean, JE, JR Goheen, TM Palmer, and TP Young. 2011. Small mammals limit tree population growth in an African savanna. Ecology 92: 1626–1636.
39. Moleele, NM, and JS Perkins. 1998. Encroaching woody plants species and boreholes: Is cattle density the main driving factor in the Olifants Drift communal grazing lands, south-eastern Botswana. Journal of Arid Environments 40: 245–253. Publisher Full Text
40. Moleele, NM, S Ringrose, W Matheson, and C Vanderpost. 2002. More woody plants? The status of bush encroachment in Botswana's grazing areas. Journal of Environmental Management 64: 3–11.
41. Muchiru, AN, D Western, and RS Reid. 2009. The impact of abandoned pastoral settlements on plant and nutrient succession in an African savanna ecosystem. Journal of Arid Environments 73: 322–331.
42. Muchiru, AN, DJ Western, and RS Reid. 2008. The role of abandoned pastoral settlements in the dynamics of African large herbivore communities. Journal of Arid Environments 72: 940–952.
43. Mwangi, E. 2007. Subdividing the commons: Distributional conflict in the transition from collective to individual property rights in Kenya's Maasailand. World Development 35: 815–834.
44. Mwangi, E, and E Ostrom. 2009. Top-down solutions: looking up from East Africa's rangelands. Environment 51: 34–44.
45. Odadi, WO, TP Young, and JB Okeyo-Owuor. 2007. The effects of wild herbivores on cattle diet in Laikipia rangeland, Kenya. Rangeland Ecology and Management 60: 179–185.

46. Odadi, WO, TP Young, and JB Okeyo-Owuor. 2009. Behavioural responses of cattle to shared foraging with wild herbivores in an East African rangeland. Applied Animal Behavior Science 116: 120–125.
47. Odadi, WO, AS Abdulrazak, MM Karachi, and TP Young. 2011. African wild ungulates compete with or facilitate cattle depending on season. Science 333: 1755–1757.
48. Okello, BD, TP Young, C Riginos, D Kelly, and T O'Connor. 2008. Short-term survival and long-term mortality of Acacia drepanolobium after a controlled burn in Laikipia, Kenya. African Journal of Ecology 46: 395–401.
49. Ogada, D, ME Gadd, RS Ostfeld, TP Young, and F Keesing. 2008. Impacts of large herbivores on bird diversity and abundance in an African savanna. Oecologia 156: 387–397.
50. Ogutu, JO, H-P Piepho, HT Dublin, N Bhola, and RS Reid. 2009. Dynamics of Mara-Serengeti ungulates in relation to land use changes. Journal of Zoology 278: 1–14.
51. Ogutu, JO, H-P Piepho, RS Reid, ME Rainy, RL Kruska, JS Worden, M Nyabenge, and NT Hobbs. 2010. Large herbivore responses to water and settlements in savannas. Ecological Monographs 80: 241–266.
52. Ogutu, JO, N Owen-Smith, H-P Piepho, and MY Said. 2011. Continuing wildlife population declines and range contraction in the Mara region of Kenya during 1977–2009. Journal of Zoology 284: 99–109.
53. 2005. Conservation and development interventions at the wildlife/livestock interface: Implications for wildlife, livestock and human health,Osofsky SA, Cleaveland S, Karesh WB, Kock MD, Nyhus PJ, Starr L, Yang A IUCN, Gland, Switzerland and Cambridge, UK.
54. Ottichilo, WK, J De Leeuw, AK Skidmore, HHT Prins, and MY Said. 2000. Population trends of large non-migratory wild herbivores and livestock in the Masai Mara ecosystem Kenya between 1977 and 1997. African Journal of Ecology 38: 202–216.
55. Porensky, LM. 2011. When edges meet: interacting edge effects in an African savanna. Journal of Ecology 99: 923–934.

56. Porensky, LM, and KE Veblen. Grasses and large herbivores reinforce landscape heterogeneity by excluding trees from ecosystem hotspots. Oecologia.

57. Pringle, RM, TP Young, and DI Rubenstein. 2007. Primary productivity and the strength of herbivore-initiated interaction cascades. PNAS 104: 193–197.

58. Prins, HHT. 1992. The pastoral road to extinction—competition between wildlife and traditional pastoralism in East Africa. Environmental Conservation 19: 117–123.

59. Prins, HHT. 2000. Competition between wildlife and livestock in Africa. In Wildlife conservation by sustainable use, ed. Prins HHT, Grootenhuis JG, Dolan TT, 51–80. Kluwer Academic Publishers, Boston.

60. Reid, RS. 2012. Savannas of our birth. University of California Press, Berkeley, CA. in press

61. Reid, RS, and JE Ellis. 1995. Impacts of pastoralists on woodlands in South Turkana, Kenya: Livestock-mediated tree recruitment. Ecological Applications 5: 978–992.

62. Riginos, C. 2009. Grass competition suppresses savanna tree growth across multiple demographic stages. Ecology 90: 335–340.

63. Riginos, C, and JB Grace. 2008. Savanna tree density, herbivores, and the herbaceous community: Bottom-up versus top-down effects. Ecology 89: 2228–2238.

64. Riginos, C, JB Grace, DJ Augustine, and TP Young. 2009. Local versus landscape-scale effects of savanna trees on grasses. Journal of Ecology 97: 1337–1345.

65. Riginos, C, and TP Young. 2007. Positive and negative effects of grasses and wild and domestic herbivores on Acacia saplings in an East African savanna. Oecologia 153: 985–995.

66. Ringrose, S, C Vanderpost, and W Matheson. 1996. The use of integrated remotely sensed and GIS data to determine causes of vegetation cover change in southern Botswana. Applied Geography 16: 225–242.

67. Sensenig, RL, M Demment, and EA Laca. 2010. Allometric scaling predicts preferences for burned patches in a guild of East African grazers. Ecology 91: 2898–2907.

68. Sitters, J, IMA Heitkonig, M Holmgren, and GSO Ojwang. 2009. Herded cattle and wild grazers partition water but share forage resources during dry years in East African savannas. Biological Conservation 142: 738–750.
69. Spalinger, DE, and NT Hobbs. 1992. Mechanisms of foraging in mammalian herbivores—new models of functional-response. American Naturalist 140: 325–348.
70. Soderstrom, B, and RS Reid. 2010. Abandoned pastoral settlements provide concentrations of resources for savanna birds. Acta Oecologica International Journal of Ecology 36: 184–190.
71. Sundaresan, SR, IR Fischhoff, HM Hartung, P Akilong, and DI Rubenstein. 2007. Habitat choice of Grevy's zebras (Equus grevyi) in Laikipia, Kenya. African Journal of Ecology 46: 359–364.
72. Sundaresan, SR, and C Riginos. 2010. Lessons learned from biodiversity conservation in the private lands of Laikipia, Kenya. Great Plains Research 20: 17–27.
73. Treydte, AC, C Riginos, and F Jeltsch. 2010. Enhanced use of beneath-canopy vegetation by grazing ungulates in African savannahs. Journal of Arid Environments 74: 1597–1603.
74. Ungar, ED, and I Noy-Meir. 1988. Herbage intake in relation to availability and sward structure: Grazing processes and optimal foraging. Journal of Applied Ecology 25: 1045–1082.
75. Veblen, KE. 2008. Livestock and wildlife effects on the successional development of a savanna landscape mosaic in East Africa. University of California, Davis. PhD Dissertation.
76. Veblen, KE. 2008. Season- and herbivore-dependent competition and facilitation in a semiarid savanna. Ecology 89: 1532–1540.
77. Veblen, KE. 2012. Savanna glade hotspots: plant community development and synergy with large herbivores. Journal of Arid Environments 78: 119–127.
78. Veblen, KE, and TP Young. 2010. Contrasting effects of cattle and wildlife on the vegetation development of a savanna landscape mosaic. Journal of Ecology 98: 993–1001.
79. Western, D, and T Dunne. 1979. Environmental aspects of settlement site decisions among pastoral Maasai. Human Ecology 7: 75–98.

80. Western, D, R Groom, and J Worden. 2009. The impact of subdivision and sedentarization of pastoral lands on wildlife in an African savanna ecosystem. Biological Conservation 142: 2538–2546.
81. Western, D, S Russell, and I Cuthill. 2009. The status of wildlife in protected areas compared to non-protected areas of Kenya. PLoS ONE 4: e6140.
82. Wigley, BJ, WJ Bond, and MT Hoffman. 2010. Thicket expansion in a South African savanna under divergent land use: Local vs. global drivers? Global Change Biology 16: 964–976.
83. Witt, ABR, and AX Nongogo. 2011. The impact of fire, and its potential role in limiting the distribution of Bryophyllum delagoense (Crassulaceae) in southern Africa. Biological Invasions 13: 125–133.
84. Woodroffe, R. 2011. Demography of a recovering African wild dog (Lycaon pictus) population. Journal of Mammalogy 92: 305–315.
85. World Initiative for Sustainable Pastoralism. 2008. Forgotten services, diminished goods: Understanding the agroecosystem of pastoralism. WISP Policy Note No. 8
86. Young, TP. 2006. Declining rural populations and the future of biodiversity: Missing the forest for the trees? Journal of International Wildlife Law & Policy 9: 319–334.
87. Young, TP, B Okello, D Kinyua, and TM Palmer. 1998. KLEE: the Kenya long-term exclosure experiment. African Journal of Range and Forage Science 14: 94–102.
88. Young, TP, TM Palmer, and ME Gadd. 2005. Competition and compensation among cattle, zebras, and elephants in a semi-arid savanna in Laikipia, Kenya. Biological Conservation 122: 351–359.
89. Young, TP, N Patridge, and A Macrae. 1995. Long-term glades in acacia bushland and their edge effects in Laikipia, Kenya. Ecological Applications 5: 97–108.
90. Zimmerman, A, N Baker, C Inskip, JDC Linnel, S Marchini, J Odden, G Rasmussen, and A Treves. 2010. Contemporary views of human-carnivore conflicts on wild ragnelands. In Wild Rangelands: Conserving wildlife while maintaining livestock in

semi-arid ecosystems, ed. Toit JT, Kock R, Deutsch JC, 129–151. Wiley-Blackwell, Oxford, UK.

91. Zwaagstra, L, Z Sharif, A Wambile, J de Leeuw, MY Said, N Johnson, J Njuki, P Ericksen, and M Herrero. 2010. An assessment of the response to the 2008 2009 drought in Kenya. A report to the European Union Delegation to the Republic of Kenya. International Livestock Research Institute (ILRI), Nairobi, Kenya.

Chapter 11

Herbicide Resistant Weeds: The Technology and Weed Management

Jamal R. Qasem[1]

[1]Department of Plant Protection, Faculty of Agriculture, University of Jordan, Amman, Jordan

INTRODUCTION

Pest resistance to control methods in general is not an isolated phenomenon but usually expected and well demonstrated when any method is repeatedly applied over a long period of time without being changed or modified in nature, structure, principals of application or formulation. All pests that growers must control in agricultural land have the capacity to become resistant to whatever tactic is used to control them [11]. It is usually expressed as a gradual adaptation or "fitness" of some individuals or populations of the targeted pest or organism to the frequently applied control methods and available conditions. This adaptation may be physical, morphological or phenological, physiological, anatomical or biochemical or could result from the interaction between any two or more of these. It may also be due to some genetic changes as mutations occur on the key site at which a specific method operates. These mutations are at least partially dominant and inherited. Traits are conferred by modifications to single nuclear genes. This indicates that the rate of resistance evolution will be driven by mutation, the intensity of selection, the dominance and relative fitness of mutations in presence or absence of the herbicide and by dispersal of resistance alleles within and between weed populations [28]. However, no proof that the herbicides cause the mutations leads to resistance [37]. However, most often resistance is controlled by a single, dominant or semi-dominant gene [38] although recessive genes control of herbicide resistant trait in natural weed populations has been also implicated in resistance to dintroanaline, while wild populations exposed to herbicide stresses for the first time may efficiently express herbicide-resistant genes.

Most weed modifications and adaptations, if not all, are advantageous to the pest, since allow its escape on time and/or place and thus avoid external hazard or threat to its existence and genetic line. Resistance therefore should not be confused with natural tolerance or low susceptibility due to a normal physiological or behavioristic property of an unselected population [23].

Organisms are varied in sensitivity, responses and thus adaptability to such conditions and in responses to any treatment or imposed external factors. Tolerance and then gradual resistance of agricultural pests to any control method or environmental stress is thus a strategy through which organisms/ or pests encounter hazards and maintain life and therefore may be applied to any method of pest or weed control including prevention, mechanical, cultural, physical, biological and chemical [30]. For example, weeds resisting soil mulch cover show some morphological and/or physical characteristics that allow penetration of the mulch layer; also, flooding of resistant species possess water impermeable seed coat or generate O_2 and reduce CO_2 penetration. Firing or flaming is resisted through presence of a hard seed coat or deeply buried regenerative propagules; certain weed species show feedback mechanisms or luxury accumulation of mineral nutrients and thus avoid toxicity; high temperature and low soil moisture harmful effects are avoided by adoption of secondary or enforced seed dormancy, while harmful effects of excessive light is avoided by some morpho-physiological alterations. Soil acidity may be encountered in the microhabitat by root exudates or selective mineral absorption and salinity by excretion of salt through different mechanisms and formation of salt glands or vacuoles or shedding salt saturated organs; microbes attack is avoided by production of repellent allelochemicals, and pests through some morpho-chemical adaptations. However, the mechanism behind tolerance or resistance is different and based on the type of target pest or the hazard imposed.

Herbicides represent one of the external factors and form a group of synthetic- plus some bio-chemicals used to suppress or kill unwanted vegetation and are a major component of pesticides. They assist in management and restoration of areas invaded by invasive species. Herbicides are a major technological tool and responsible, in part, for an agricultural revolution and increase in food production in the last few decades. However, at present this technology faces radical changes in effectiveness under field conditions that lead in different cases to failure of weed control operation due to continued development of weed tolerance/resistance and evolution and limitations in the herbicide industry and development.

AGRICULTURE PRACTICES AND WEED EVOLUTION

General weed control methods (tillage, hoeing, hand weeding, flooding, cuttings or mowing, flaming, use of general herbicides) are all nonselective and usually applied to a composite weed species or vegetation of inter and intra-specific variations in richness, morphology, growth habit and responses. Each species may adapt, or not, to any of these methods. Since weeds are widely different in mechanisms by which they encounter hazards they are exposed to, they are different in plasticity and responses. With continued use of a single control method for a long period of time, species migrate, flourish or die. Flourishing species gradually became better fit and adapted, and increase in number and population size in absence of others. The only surviving individuals are those possessing rare single gene mutations and evolved resistance will be monogenic, resulting in a large change in the resistance phenotype. However, when doses are lower and selection acts within the range of standing genetic variation, polygenic responses will be possible and resistance will evolve by a gradual change in the mean susceptibility of the population [28]. On the other hand, population of not or less adapted individuals, decline in growth and number until greatly suppressed, limited and may become extinct. Therefore, with continuous dependence on a single method of weed control, a weed population is usually shifting toward better adapted species or individuals that cope well with existing control measures and new conditions. Self-thinning of a weed population is continued toward complete tolerance to employed control measures. Therefore, weeds adapted to mowing tend to grow short, in a rosette form, creeping above the soil surface or show high plasticity and softness of aerial parts and stems and become difficult to mow and also escape hand weeding. Deep rooted weed species are difficult to pull out even by soil tillers. Seasonal dormancy and shifts in the weed population in the growing season is well recognized for certain weed species such as *Senecio vulgaris* [29; 37], while physiological adaptation of *Echinochloa crusi-galli* and *Cyperus rotundus* to flooding conditions and the role of Alcohol dehydroginase enzyme (Adh) in *E. crusi-galli* is well documented [5; 14]. Similar adaptations of *Cirsium arvense* ecotypes to temperature variations [43] and *Typha anguistifolia* and*Typha latifolia* genetic and clonal variations [27; 40] have also been reported. In this regard, it is important to differentiate between tolerance and resistance of weeds to herbicides. Tolerance is the inherited ability of a species to survive and reproduce after herbicide treatment; it refers to the natural variability to herbicides and exists within individuals of a species and quickly evolves. It usually refers to relatively minor or gradual differences in intraspecific variability. Resistance is the inherited ability of a plant or a biotype to survive and reproduce following exposure to a dose of herbicide that is normally lethal to wild type [16; 23; 30; 37]. Therefore, it is a decreased response of a population of weed to herbicides as a result of their application. However, both terms sometimes are misused or used interchangeably.

Tolerant weed species are less harmed by herbicides; they exhibit a certain degree of avoidance or adaptation strategy that allows recovery and thus escape control measures. They may respond by timing stomata closure or having sunken pores or stomata, thick waxy cutical on upper leaf surface, encased growing points or some biochemical, physiological or anatomical properties better developed by time until they become best fit and adapted to applied herbicides and become thereafter resistant. This, however, leads to gradual but radical changes in the weed population composition and distribution spectrum at which resistant individuals or certain weed species increased and dominate and susceptible ones are reduced and replaced. Adaptation or exclusion of the less tolerant species depends on performance of these by time. Generally a weed population becomes rich in individuals and poor in species with the continuous use of the same herbicide or different herbicides of similar mode/mechanism of action. This shift does not however, reflect better competitiveness or higher regenerative ability but most likely due to absence of sensitive highly competing species or forms that allow resistant individuals to utilize more resources [9; 22].

In cultivated fields, associating weeds bear more resemblance to crop plants in morphology, physiology and responses to control measures and other agricultural practices in general. They mimic crops from sowing and germination until harvest. Since herbicides used on crop plants are selective, weeds respond by exhibiting similar morphology, physiology and biochemistry as crop plants to avoid hazards. However, weeds derived from crop plants as hybrids, crop relatives or wild-weedy forms are better fit to such conditions than others. Weed-crop associations also exist between weed species of different taxa from crop plants. In this case, the longer the use of the same herbicide/s, the greater the close association between crops and certain well performed weed species that later transfer into adapted weed races. Crop relative weeds however, are of great potential to intra- and inter- gene exchange and efficient mating system among themselves and with crops, thus become best adapted and more difficult to control.

SELECTION PRESSURE AND WEED RACES

With continuous use of the same agricultural practice/s, interspecies selection occurs and plant species are gradually purified (intraspecific selection) by time until they become best adapted. Since all control measures including herbicides aim to eliminate weeds without causing injury to crop plants, weeds respond by developing mechanism/s allowing escape of chemical hazards. Under such conditions, sensitive individuals are first limited or disappear. Tolerant individuals increase in number and accumulate tolerance until they become resistant.

Therefore, a resistant population of any weed species is exposed to long-term selection pressure through which it is purified and performs well under prevailing conditions in absence of sensitive weed species. With continuous exposure to herbicide pressure, a population of resistance is usually developed.

Weeds tend to avoid herbicide toxicity by changing normal growth habits, or exhibiting some phenological (such as changes in germination patterns), physical and/or physiological changes through which they adjust emergence time, external appearance or physiology. These however, are inherited traits that allow plants to survive herbicide treatments. One best adaptation is that of weeds similar to crop plants in most or all growth aspects. These form weed races similar to crop plants and well adapted to their habitats. Among reported weed races are *Camelina sativa* to flax crop, *Echinochloa crus-galli var. Oryzicola* that associate with rice and the weedy wild rice or red rice in India and east-south Africa [8; 20]. All are genetically irrelevant to crop plants. However, in some cases weed races are of the same botanical family or belong to the same crop species. This kind of association leads to development of "crop- races" that possess weedy characters very well adapted to cultural practices; they are similar to crop plants in most growth aspects and difficult to control by herbicides or other control methods including hand weeding. They take an advantage from conditions under which crop plants are growing until they become difficult to leave their habitats or even become dependent on crop plants in their growth and environment. These weeds are specialized to certain crop plants or cultivars. Moreover, many genetically related species can exchange genes with crop individuals and mimic crops. It can be concluded that any agricultural practice exerts selection pressure and may become troublesome to farmers when repeatedly applied for a long period. Its positive impact on crop growth and productivity is usually negated with time until it becomes a real trouble. Its residual negative effects may not possible to overcome for a long period after abandonment.

FIELD EVIDENCE OF WEED RESISTANCE AND HERBICIDE RESISTANCE PROTOCOL

In the field all growth patterns and distribution of weed species may be observed. Some species grow in colonies, in certain growth patterns, forming an ecological niche, sporadically distributed, or randomly scattered within crop plants. Certain species are dominant while others show moderate growth or are suppressed while some grow vigorous or have limited growth and short stature. This however, depends on the microhabitat and place they occupy in the field and their performance. Under intense cultivation and thick crop stands, individuals of certain weed species express phenotypic plasticity (phenotypes) at which they change/modify their appearance, reduce or drop lower branches and thus lateral growth, elongate and increase cell divisions, overtopping crop plants and trapping light, although some shade tolerant species perform better under such conditions. Phenotypic plasticity modifying the mode of growth and energy allocation in response to environmental changes is considered to be important adaptive mechanism. These phenological variations can be easily observed among different weed species. Uniform application of herbicides in the field should equally affect all individuals of a single weed species. When herbicides are best timed and properly applied they should yield similar mode of action on species individuals. While differences in influence of a herbicide on different weed species is expected, hence differences in taxonomy, morphology, physiology and biochemistry, but such differences among individuals of a single species should have resulted from some morphogenetic or other variations within the same or different populations of that species. Certain individuals are totally killed, others less injured and some escape control unharmed. When the same herbicide or herbicides of the same mechanism of action are used, it becomes clearer that previously less or unaffected individuals should exhibit similar responses as were first shown. Gradually these individuals increase in number and growth until they dominate the site with continuous use of the same herbicide or its analogues while sensitive individuals are suppressed or removed. This however, takes a relatively long time for the population to shift from susceptible to complete resistant and depends on herbicide, environment and plant factors. These are positive signs on possible herbicide-resistance development in the field. If less affected or unharmed individuals in the first herbicide application are killed or severely injured in repeated treatments then there should be another cause of escape or partial control at first application and herbicide resistance should be then excluded. On the other hand, unharmed individuals may also tolerate higher application rates. Therefore, farmers should keep observing changes in the weed population as long as the herbicides are in use. They must get familiarized with weed species, populations and densities at pre- and post- herbicide treatments, comparing weed growth, performance and densities and recording any changes in populations thereafter. Less or unharmed individuals of any species should be followed up throughout subsequent applications of the same herbicide or herbicides of similar mode of action.

Sometimes partial effect or failure of the applied herbicide to control certain weed species or individual weeds in the first application may be thought as due to wrong calibration, misapplication, incomplete coverage treatment by a general herbicide or unsprayed gaps resulting from low sprayer boom during spray, unfavorable weather conditions, improper timing of herbicide application, and weed flushes after application of a non-resisted herbicide [16]. This could be easily judged in the repeated application to these species or individuals. When the herbicide failed to control these for the second time or at higher rates then resistance may be underway. With continued use of the same herbicide for different times, resistant individuals aggregate forming irregular patches while other weeds are controlled. A patch of uncontrolled weeds starts spreading and healthy weeds are mixed with uncontrolled weeds of the same species (Fig. 1).

Therefore irregularly shaped patches of a single weed species in the field are an indicator of herbicide resistance, especially when:

- There are no other apparent application problems.
- Other weed species on the herbicide label are effectively controlled.
- Field history indicates extensive use of the same herbicide or herbicides of the same mechanism of action.
- No or minimal herbicide symptoms appear on the single uncontrolled weed species.

- There has been a previous failure to control the same species or population in the same field with the same herbicide or with herbicides of the same site of action.

However, the rate at which a resistant weed population is selected depends on the number and frequency of herbicide applications it receives, the size of the population and its genetic diversity, and characteristics of the herbicide target site. Resistance buildup is accelerated when the management of crops does not include different weed control methods that limit herbicide use. In addition, this may be greatly enhanced in conservation or zero tillage because weeds are not killed by mechanical disturbance and general herbicides.

INTERACTION BETWEEN ENVIRONMENT AND GENETICS

Growth and productivity of any plant species are mainly influenced by genetics, ecology and their interactions. Weeds are different from crops in their responses to both factors. They are more flexible and thus better responsive and adapted to extremes in environmental conditions such as high temperature, freezing, excessive light, salinity, drought, etc. Tolerance of weeds and better responses are mainly due to better and rapid interaction between environment and genetics compared to crop plants. In addition, the long term breeding and selection pressure imposed on crop plants has led to selection of less adapted species or cultivars that are highly sensitive to ecological stresses and deficient in certain characteristics that offer protection or defense mechanisms against unfavorable environment. Weed fitness in natural habitats and their rapid responses to the changing environment allow evolution of weed ecotypes, genotypes, biotypes or phenotypes. Some of the basic differences in the definitions of pest resistance depend on these terms. The basic unit of plant classifications is the "species" that is defined as a group of individuals displaying common characteristics and having the ability to mate and produce fully viable progeny. A species usually consists of several to many populations. A population is a group of organisms within a species that co-exist in time and space [35; 36] and share a distinct range of genetic variations. While a genotype is the sum of the genetic coding or the genome of an individual, a biotype may not be coincident with genotype as an individual has many genes. Certain genes may be expressed or unexpressed and not pertain to the phenotype associated with the biotype. A biotype is a phenotype that consistently expresses or exhibits a specific trait or set of traits; it represents a group of individuals or a population within a species with a distinctive genetic variation of biochemical or morphological traits. Phenotype refers to the physiological and morphological profile of the expressed gene in an individual [42]. A single genotype can produce different phenotypes in response to environmental conditions and the fundamental properties of organisms are known as phenotypic plasticity. The epigenetic change is thus reflecting the alteration of phenotype (morphological or biochemical) without change in either the coding sequence of a gene or the upstream promoter region. Therefore biotypes within the same species may be developed due to this interaction. On the other hand, ecotype is a population within a species that has developed distinctive morphological or physiological characters (herbicide resistance) in response to a specific environment and persists when individuals are moved to a different environment. Ecotypes are of different germination and growth optima for the same environmental factor and phenotypes may be emerged and observed in weed populations. These alter their morphological features in response to certain prevailing environmental conditions which aim at protection of their individuals against unfavorable ecological stresses. Somatic polymorphism of certain weed species is well recognized and expressed as seed polymorphism of different morphological or physiological requirements for germination on different parts of the same weed individual. These however, are somatic rather than genetically based differences.

Figure 1: Three resistant weed species (a, b, c) to glyphosate herbicide at different growth stages and spray times. (a). *Conyza canadensis* resistant to glyphosate until harvest stage of wheat. Source http://www.sciencephoto.com/media/ courtesy of the Montana State University (b). A field infested by suspected glyphosate- resistant *Kochia*, after the field was sprayed with three applications of glyphosate. Photo181407/enlarge Southern Agricultural Research Center. By Dillon Tabish, 08-11-12.Available at: http://www.flatheadbeacon.com/articles/article/scientists_discover_possible_herbicide_resistant_weed_in_montana/29184 (c). Palmir Amaranth (*Amaranthus palmeri*) resistance to glyphosate in corn at early growth. Source: E. Larson, April 21st, 2011.Availableat:http:// www.mississippi-crops.com/2011/04/21/how -to-deal-with-glyphosate-resistance-and- weed-issues-in-corn/.

HERBICIDE RESISTANCE AND CROP RELATIVE WEEDS

Crop relative weeds are usually derived from the same species of crop plants and thus are genetically related. Most crop species have wild relatives and can interact with them under field conditions. Examples are radish, carrots, vetch, celery, lettuce, fennel, eggplants, wheat, barley, oat, etc. In addition, crop plants which are domesticated from wild forms possess a high degree of compatibility with crops. These are referred to as wild and weedy relatives, in spite of the fact that all species are related because their cells can read a common genetic code [15]. Crop weedy relatives are genetically compatible with crop plants and easily exchange genes. The emerged hybrids may become noxious weeds with certain weedy characteristics derived from both crop plants and wild forms. They could exhibit a certain degree of dormancy that is usually weak or absent in its parents and possess other weed traits making them difficult to control. These new generations have the ability to resist environmental hazards much better than parents and can exist and dominate in both productive and unproductive habitats. These are of a high genetic plasticity allowing their individuals to adapt to extensive herbicide applications and thus resist chemical treatments. Crop-weed crossed forms can easily exchange genes with crop plants as well as with weedy relatives and therefore are becoming troublesome weeds in fields with genetically modified crops.

GENE FLOW POTENTIAL WITH WILD/ WEEDY RELATIVES OF WORLD CROPS

In nature, genetic information is transferred between different individuals, populations, and generations (to progeny) and across spatial dimensions [2; 15]. This phenomenon, known as gene flow, serves as a mechanism to maintain the biological diversity that helps to ensure long-term survival of populations and species in various environments.

Gene flow is a critical determinant of population genetic structure, playing an important role in both evolutionary and applied plant population genetics [12]. It is also known as 'migration' [13] or admixture [1] and can be defined as the movement of genes between populations of a species and between these populations and inter-fertile relatives [39; 41], conferring new traits, the biophysical characteristics of the organism to individuals of the recipient population [34].

Gene flow could occur through dispersal of pollen (via outcrossing between sexually compatible individuals within or among populations) or seeds (via seed dispersal), or vegetative parts capable of clonal propagation [34; 41]. Pollen dispersal is the typical method for such exchange of genetic information [15] and pollinating visitors or other agents including wind, animal, water current and other factors could play a significant role in this issue. This happens by cross-pollination (hybridization), that is, the pollination of members of one population or genetic pool with that of another [34]. These are natural and ordinary phenomena that occur in conventional as well as genetically modified crops.

Movement of pollen away from its site of production can result in true gene flow only if (1) the pollen first effects fertilization to form seeds, and (2) seeds germinate, produce plants that express the gene (i.e., are not silen8ced), and are able to reproduce [15]. Gene flow can be from crop to crop or landrace, from crop to wild relative, and even from wild relative to crop plant [34]. Spread of this phenomenon would lead to radical changes in vegetation composition and weed ecological distribution and their economic significance.

However, two types of gene flow are known; horizontal and vertical. Stewart [39] showed that 'horizontal' gene flow is the movement of genes between disparate, unrelated species, such as between plants and microbes while horizontal gene flow is more theoretic.

Among the world's 180 most damaging weeds, however, cause 90% of all crop losses, only five groups (related weeds of rice, sorghum, rape seed, sugarcane, and oats) are sexually compatible with the most important crops (Table 1). This fact emphasizes that the number of weed-crop crosses likely to lead to extremely troublesome or unmanageable problems is small.

Weed crosses with herbicide-tolerant biotech crops are likely to be favored in some agricultural fields where the herbicide is used. In areas where little or no herbicide is applied (e.g., native lands), the weed–biotech crop crosses will not be favored [15]. Self-pollinating crops are considered of low risk in terms of gene flow to weeds. Roundup Ready, Clearfield, or Liberty Link canola, in contrast, could pollinate nearby herbicide-susceptible canola as well as weedy canola relatives, resulting in volunteer canola plants and weeds that may be resistant to several herbicide families [38]. However, several pieces of evidence clearly show an escape of weedy transgene from fields via seed flow and this escape occurs via man-mediated long-distance dispersal events [4]. Other results revealed that development of weed resistance via selection pressure from repeated herbicide applications in herbicide resistant crops (in the absence of gene flow), often poses greater risks than that from gene flow to related weed species [15].

Table 1: Examples of some important food crops and their sexually compatible weed species

Rank	Crop	Scientific Name	Related weeds: sexually compatible with crops
1	Wheat	Triticum aestivum Triticum durum	T.aestivum Aegilops cylindrical A. tauschii A. triumcialis Agropyron spp
2	Rice	Oryza sativa Oryza glaberrima	O. sativa O. glaberrima O. barthii O. longistaminata O. rufipogon O. punctata
3	Maize	Zea mays	Z. mays ssp Mexicana
4	Soybean	Glycine max	G. soya
5	Barley	Hordeum vulgare	H. spontaneum
6	Sorghum	Sorghum bicolor	S. bicolor S. almum S. halepense S. propinguum S. sudanense
7	Canola	Brassica napus, B. rapa, B. juncea	B. napus, B. rapa, B. nigra
8	Sunflower	Helianthus annus	Helianthus annus

[i] Source: Different references

In this regard, biotech crops conferring stress tolerance (e.g., to water deficits, diseases, insects, salt stress, or nutritional deficiencies) may need more scrutiny because their crosses with weedy relatives may impart selective advantages in both agricultural and nonagricultural areas. Thus, some traits obtained from biotech crops could theoretically facilitate development into problematic weedy or wild species [15].

The economic consequences due to gene flow from biotech crops will primarily impact the agricultural fields in which those crops are grown, but potentially could impact natural areas given the proper rare combination of sexually compatible relatives, favorable environment, and reproductive/fitness advantages. As an example, rice grown in tropical countries may be relatively more prone to such processes because of the substantial populations of its wild/weedy relatives that grow naturally in or adjacent to the rice-producing areas [8; 26].

Crop-wild hybridization may also create genotypes with the potential to displace parental taxa in new environments [7]. However, the most important variable affecting gene flow is the degree of relatedness and distance between the crop and the weed, because gene flow is only possible if close relatives are growing near the crop. As a result the possibility of gene flow depends mainly on presence of wild or weedy relatives [11]. Transgene (s) transfer may have unpredictable and out of control ecological impacts under intensive cultivation of biotech crops [25]. While different crops can exchange genes with wild relatives, gene escape to wild or weedy relatives and its ecological impacts are outrated. The ecological consequences of gene flow however, depends on the amount of transgenes moved out to a wild population and the genetically modified traits and whether they have an evolutionary advantage under natural selection pressure or not and if enhanced fitness of wild and weedy relatives then the transgene followed by gene flow would persist and spread rapidly in the population of wild relatives through introgression, invade a new area and outcompete other individuals under natural conditions [24]. Weeds receiving transgenes will continue to evolve when exposed to selection pressure and it becomes nearly impossible to move them out from the environments if they can persist and spread in the populations.

TRANSGENIC CROPS AND WEED EVOLUTION

The development of crops that are resistant to herbicides is a relatively new technology aimed to improve weed control in agricultural land. Herbicide-resistant crops can be created by standard methods of plant breeding, but the use of genetic engineering techniques is more usual. Herbicide-resistant crops are made resistant by either transgene technology or by selection in cell or tissue culture for mutations that confer herbicide resistance [10]. Glyphosate and glufosinate are herbicides most used in this regard. For example, soybean, corn, cotton, sugar beet, and canola are available as glyphosate- resistant cultivars and some are now widely planted in different countries. Importance of genetically engineered crops is to:

- Develop crops more tolerant/resistant to herbicides and thus increase herbicides uses and selectivity.
- Eliminate possible injury effects of soil persistent herbicides to crop plants.
- Increase options for weed control when the number of herbicides is limited, such as in minor crops.
- Effective control of certain difficult weed species and widening of weed control spectrum
- Achieve more effective weed control
- Increase bio-safety and enhance better eco-friendly use of new and less toxic herbicides
- May be more cost- effective weed control method

However, public concern about the impact of genetically modified crops on the natural environment encouraged more studies on this aspect in the last few years. Among the possible impacts, the 'escape' of the transgene, either through dispersal of the crop plant outside the agricultural area or through hybridization with wild relatives and thus increase the possibility of "weediness" [41].

In the majority of instances, there is a very low probability that an approved biotech crop introduction could create an environmental risk different from that of a nonbiotech version of the same crop. This however, does not lessen the serious concerns about possible consequences of the escape of transgenes into the environment [41]. Examples of the risks mentioned in the context of gene flow from genetically modified plants are: i) new emerged weeds resulting from an escape by the crop itself; ii) super weeds resulted by hybridization of a (wild/weedy) species with the transgenic crop; iii) genetic erosion (loss of original diversity of wild relatives). To date, all instances of weeds becoming resistant have resulted from the weed evolving its own biochemical mechanism and not by acquiring genes for resistance from the crop. However, in some cases it would be possible for the herbicide resistance gene to flow from the crop to the weed [11].

Possible consequences of hybridization and introgression depend on the plant, gene, trait, and ecological factors [39]. In the case where transgenes might be introgressed into "weedy wild relatives", there are concerns about exacerbating "weediness" traits or even the disruption of natural ecosystems. Therefore, to assess the risk of gene flow it needs to be examined not only the probability of genes moving between plants, but how possible is it for the new plants to survive [39].

In general, people ideally would like to minimize or prevent gene flow from transgenic organisms to weedy wild relatives or to places where extensive crop breeding takes place [39]. Three approaches to gene flow mitigation are possible [3].The first is by keeping the genetic modification out of the pollen, preventing the formation of pollen, and keeping the pollen inside the flower. It requires transplastomic plants hence the modified DNA is not situated in the cell's nucleus but is present in plastids, which are cellular compartments outside the nucleus. The second approach relies on male sterile plants unable to produce functioning flowers and therefore cannot release viable pollen. Cytoplasmic male sterile plants are known to produce higher yields. The third approach works by preventing the flowers from opening "cleistogamy" that occurs naturally in some plants. Cleistogamous plants produce flowers which either open only partly or not at all.

However, herbicide-resistant genes have no ecological significance in places where the corresponding herbicide is not used. When paired with a gene that might have an effect in a natural ecosystem, there is a potential problem with gene flow. Repeated application of the herbicide (especially general herbicides) would select for and protect crosses and backcrosses, increasing the possibility of successful gene flow to wild, related species [10].

WEED CONTROL SPECTRUM OF SELECTIVE HERBICIDES AND POPULATION SHIFTS

Some plants are genetically tolerant to certain herbicides while others have evolved resistance after repeated exposure to an herbicide. Tolerant and resistant plants usually degrade or metabolize the chemical to nonphytotoxic substances. In some cases of resistance, such as with triazine herbicides, the herbicide does not reach the key site in treated plants. Although tolerance and resistance are common, herbicide selectivity among plants is often conditional; thus it depends on plant, herbicide and environment factors.

Some of the factors that influence herbicide selectivity are as follows:

- Physiological or biochemical tolerance to the herbicide
- Herbicide application rate
- Time of application
- Herbicide formulations and surfactants used.
- Growth stage of weed and crop or other plant development
- Weather patterns (temperature, light, wind, rain, etc.)
- Variation in microenvironment or micro- topography
- Variation in resource level
- Soil type and pH

Many of the principles and practices of how herbicides used or applied to attain selective chemical and effective weed control are important. These involve the role of plant morphology and physiology, chemical properties, and environmental factors [31]. Herbicide selectivity in one way or another is in direct link with herbicide resistance. Crops are resistant to herbicides selectively used to kill weeds. Even with repeated treatment, crop plants can resist or tolerate higher rates of selective applied herbicide or repeated treatments. This depends on some level of tolerance/resistance higher in crop plants compared with weeds for that specific herbicide or herbicide group. For example, Syrian marjoram (*Origanum syriacum*) was found to withstand up to 4 times higher rates of oxadiazon and oxyfluorfen herbicides either applied on foliage parts or through the soil [32; 33]. Certainly many factors have an important role in giving a resistant value for crop plants. Some of these are listed below:

Plant Factors and Herbicide Selectivity

Plant factors that influence the way weeds and crops respond to herbicides are genetic inheritance, age, growth rate, morphology, growth form and anatomy, and physiological and biochemical processes. The most effective use of herbicides results from considering these factors when selecting an herbicide or application method.

Plant Age and Growth Rate

Weed seedlings or young plants are usually killed more easily than large or mature vegetation. In addition, some preemergence herbicides that suppress seed germination are often not effective when used to control larger, better established plants. Plants that are growing rapidly or in shaded places generally are more susceptible to herbicides than are plants of slow growth or unshaded.

Morphology

The morphology or growth habit of plants can determine the degree of sensitivity to some herbicides. Morphological differences in root structure, location of growing points, and leaf properties between crops or other desirable plants and weeds can determine the selectivity pattern of some herbicides. Annual weeds in a perennial crop, meadow, or pasture usually can be controlled by herbicides because of their different root distribution and structure compared to those of perennial plants. For example, perennial crops such as alfalfa can recover from moderate contact herbicide injury to foliage whereas annual weeds, because of their small size and shallow root system, will be killed by the same herbicide application.

The meristematic regions of most grasses, such as cereal crops and grassy weeds, are located at the base of the plant or even below the soil surface. The growing points are protected from herbicide exposure by the foliage or soil that surrounds them. Thus, herbicide that contacts only foliage may injure some leaves but will not typically impair the ability of the plant to grow. In contrast, most dicot plants have their meristems exposed at shoot tips and leaf axils. For this reason, these plants are more susceptible than grasses to foliage-applied herbicides, especially of contact action.

Leaf properties of some plants can impart selectivity to certain herbicides, while other plants are effectively controlled. Spray droplets do not adhere well to the surfaces of narrow, upright, waxy leaves that characterize many monocot plants like cereals, onion, and most grasses. Thus, spray droplets do not adequately cover such leaves following herbicide application and the effect of the herbicide is reduced. In contrast, dicot plants have relatively wide leaves that are usually horizontal to the main stem. Leaves of dicot plants, therefore, intercept more spray solution than leaves of grasses and spray droplets spread more evenly over dicot foliage. Herbicide effectiveness is best when spray interception and coverage are greatest and with use of surfactants. However, ecological factors and geographical regions under which weeds are growing have significant influence on herbicide selectivity and rates of applications since they affect or modify weeds morphology and internal anatomy.

Physiological and Biochemical Processes

Plant physiology influences herbicide passage after its application. This process is called "absorption". The extent of herbicide movement in a plant- "translocation"- after it has been absorbed is also a physiological process. Both absorption and translocation are important processes governing herbicide activity and vary markedly among plant species. Generally, plant species that readily absorb and translocate herbicides are most easily killed.

Biochemical and biophysical processes are also important plant factors determining herbicide selectivity. Herbicide adsorption can be responsible for differential herbicide susceptibility among plant species. During this process an herbicide is bound so tightly by cellular constituents (usually cell walls) that it cannot be translocated readily and thus is inactivated. Membrane stability is another biochemical/ biophysical process that results in herbicide selectivity among plants. In this case, the cell membranes of tolerant plants can withstand the disruptive action of the herbicide. The ability of carrot to withstand the toxicity of certain oils is an example of this form of herbicide selectivity.

Genetic Inheritance

Plant species within a genus usually respond to herbicides in a similar manner, while responses to herbicides by plants in different genera often vary. The reason is that plants with similar taxonomic traits often have similar morphogenetic and enzymatic components. Thus, crops and weeds that belong to the same genera are usually susceptible to the same herbicides and are similarly affected since they have similar biochemistry. This rule is not absolute, however, because varieties of many crops are known to respond differently to the same herbicide and weeds usually adopt different mechanisms of herbicide resistance while crop plants have lost many of their traits in breeding programs that present in wild relatives.

HERBICIDES AND EDAPHIC FACTORS

Soil factors affect herbicide performance and their effectiveness. These including soil-organic matter content, microorganism populations, soil water table and moisture content and soil pH. Organic matter acts through adsorption and release of chemical molecules. Certain herbicides are tightly adsorbed on soil particles and thus become unavailable to weeds. These molecules may be totally inactivated upon their release. Therefore weed control may be complete or not based on the amount of the herbicide adsorbed and whether the held amount on soil colloids is compensated or not before applied. The higher the percentage of organic matter and clay particles, the greater the adsorption in amount and time of herbicide molecules and the lower the herbicide activity and *vice versa*. This requires that some operations should be well managed when soil applied herbicides are used including their incorporation or placement in/on the soil.

Activity of soil microorganisms is another factor affecting activity of soil- applied herbicides and persistence. Microorganisms may degrade herbicide molecules and feed on organic herbicides. In general, favorable soil factors to microorganism populations stimulate their activity and thus rapid herbicide degradation. Therefore, soil-microbe population is an important factor in increasing or decreasing herbicide persistence and weed control duration.

Soil water also affects herbicide activity and performance. When high amounts of soil water are available or at high soil water levels, herbicide molecules may by hydrated. On the other hand, moisture is necessary to transfer herbicide molecules into the root system and then translocate these upward to vegetative parts through the xylem.

Soil pH affects cation exchange capacity of soil particles. Salt or mineral forms of certain herbicides may interact with soil particles under these conditions by exchanging cations or anions and thus lead to breakdown of herbicide molecules and inactivation.

All above soil factors and others such as soil- root temperature and soil mechanical properties can affect herbicide activity and performance and their effectiveness in controlling weed species and herbicide selectivity. Weeds may become adapted to certain soil conditions, escape control operations and lead to dominance of well adapted species or populations.

WEED RESISTANCE AND DORMANCY, AVOIDANCE AND WEED DENSITY

Dormancy is the state at which seeds in the soil or buds are not germinating or growing due to external conditions exert influences on physiological and biochemical internal processes including enzymes activities, food transport to embryo and metabolism. This state is keeping seeds or buds safe until the cause of dormancy is over. This behavior is important to maintain genetic line and continuity of the species in changeable environment. Under conditions of herbicide application, some of these chemicals are absorbed by seeds or dormant buds while others are not. These result differences in germination, emergence and growth patterns of different weed species. However, some herbicides may stimulate seed germination while others inhibit this process or even kill seed embryo. Differences also exist in hardness and permeability of seed coat of different weed species at which species of Chenopodiaceae and Fabaceae are good examples. These characters cause differences in germination and growth of seedlings and may confer another cause of herbicide resistance. Avoidance of herbicide toxicity may result from seed interring into dormancy and not further responding to the applied herbicide with no absorption or translocation of the herbicide into the embryo. In addition, herbicide molecules may be deactivated or degraded inside the seed itself by some oxidative enzymes or may bound into certain constituent inside the seed.

On the other hand, stimulation of weed seeds to germinate using certain herbicides also exist and allows higher seedlings emergence and partitioning of herbicide molecules among individuals of weed species. Division of herbicide molecules among high number of emerged seedlings would further diluted herbicide inside weed plants.

All above mentioned factors should be considered when herbicide-resistance is discussed. These may cause great differences in weed growth patterns and distribution in the field.

WEED RESISTANCE UPDATES AND RESISTANCE MECHANISMS

With continued dependence on herbicides for weed control and with the absence of other methods and herbicide rotation, the resistance problem is extenuated and the number of resistant weed species and biotypes is dramatically increased. At present, the reported herbicide resistant weeds are approaching 393 (species and their biotypes). These represent 211 species (124 dicots and 87 monocots) and detected from over 680,000 fields [21; 44] reported from 61 countries from all over the globe. However, the highest number of resistant species was reported from the advanced countries indicating efficient and rapid detection with available technology to diagnose, discover and deal with this issue. However, the highest number of weeds reported resist the main three groups of herbicides based on site of action including; the ALS (127 weeds), Photosystem II (69) and the ACCase (42) inhibitors. The highest number of weed resistant species and biotypes came from the USA (141), Australia (61) and Canada (58). Most numbers of resistant species belong to the families Poaceae, Asteraceae and Amaranthaceae and most frequently mentioned are genera of *Amaranthus* (30 times and 11 species),*Echinochloa* (23 times and 6 species), *Lolium* (20 times and

4 species), *Alopecurus* (12 times and 3 species), *Avena* (11 times and 3 species), *Bromus* (11 times and 5 species), *Conyza* (10 times and 3 species), *Setaria* (9 times and 5 species), *Poa* (8 times and one species), *Ambrosia* (7 times and 2 species), *Digitaria* (6 times and 4 species), *Phalaris* (6 times and 3 species), *Hordeum* (5 times and 2 species) and *Sorghum* (6 times and 3 species). Most are of the grass family usually exhibiting distinct morphological features allowing wide dispersal and escape of herbicide treatment such as encased growing points, vertical leaf arrangement and thick waxy cuticle that reduce herbicide penetration and lead to herbicide droplets bouncing off leaves. Other genera reported are characterized by their prolific seed production and/ or seed polymorphism. All above mentioned genera however, showed multiple resistance to different herbicides groups. Most resisted are herbicides widely and repeatedly used including: glyphosate, paraquat, atrazine and 2,4-D and others used in fields cultivated by genetically modified crops. Some recently developed herbicides are also resisted including chlorsulfuron and sufonylurea group. This phenomenon demonstrates that the herbicide industry and development is far behind weed evolution. On the other hand, weed species and biotypes showing multiple resistance are most common and some are among the world's worst weeds [19] including: *Amaranthus* spp.,*Echinochloa* spp., *Avena* spp. and *Chenopodium album* characterized by their polymorphic seed production and phenotypic plasticity. This reflects a great ability to maintain and exhibit high plasticity and possess various mechanisms of herbicide resistance.

The precise molecular mechanism of resistance varies with different plants, but in general plants resist herbicides in one of the following ways:

- Avoiding the herbicide by not absorbing it or, if absorbed, the weed compartmentalizing it away from its target site.
- Reducing the uptake or herbicide uptake is not enough to injure the weed or reach lethal level.
- Changing the structure of the target site of the herbicide so the plant is no longer sensitive
- Reduce herbicide translocation to the key site or binding it into certain plant constituent
- Sequestration by complete physical removal of the herbicide from the key site

- Target site mutation and changes in structure lead to insensitive plants and failure herbicide binding.
- Deactivating the herbicide by chemical alteration or herbicide metabolism before reaching target site

However, resistance mechanisms through which different weed species resist herbicide treatments are many and varied but most are physio-chemically based (Table 2).

Table 2: Herbicide resistant weeds summary table (Thursday, November 08, 2012)

Herbicide Group	Site of Action	HRAC Group
ALS inhibitors	Inhibition of acetolactate synthase ALS (acetohydroxyacid synthase AHAS)	B
Photosystem II inhibitors	Inhibition of photosynthesis at photosystem II	C1
ACCase inhibitors	Inhibition of acetyl CoA carboxylase (ACCase)	A
Synthetic Auxins	Synthetic auxins (action like indoleacetic acid)	O
Bipyridiliums	Photosystem-I-electron diversion	D
Glycines	Inhibition of EPSP synthase	G
Ureas and amides	Inhibition of photosynthesis at photosystem II	C2
Dinitroanilines and others	Microtubule assembly inhibition	K1
Thiocarbamates and others	Inhibition of lipid synthesis - not ACCase inhibition	N
PPO inhibitors	Inhibition of protoporphyrinogen oxidase (PPO)	E
Triazoles, ureas, isoxazolidiones	Bleaching: Inhibition of carotenoid biosynthesis (unknown target)	F3
Nitriles and others	Inhibition of photosynthesis at photosystem II	C3
Chloroacetamides and others	Inhibition of cell division (Inhibition of very long chain fatty acids)	K3
Carotenoid biosynthesis inhibitors	Bleaching: Inhibition of carotenoid biosynthesis at the phytoene desaturase step (PDS)	F1

Glutamine synthase inhibitors	Inhibition of glutamine synthetase	H
Arylaminopropionic acids	Unknown	Z
Unknown	Unknown	Z
4-HPPD inhibitors	Bleaching: Inhibition of 4-hydroxyphenyl-pyruvate-dioxygenase (4-HPPD)	F2
Mitosis inhibitors	Inhibition of mitosis / microtubule polymerization inhibitor	K2
Cellulose inhibitors	Inhibition of cell wall (cellulose) synthesis	L

[i] - Source: 21; Updated: November, 2012

FACTORS ENHANCING HERBICIDE RESISTANCE

All natural weed populations, regardless of the application of any herbicide, probably contain biotypes that resist herbicides. Repeated application of an herbicide exposes the weed population to a selection pressure which may lead to an increase in the number of surviving resistant individuals in the population. As a consequence, the resistant weed population may increase to a level that adequate weed control cannot be achieved by the application of that herbicide [18]. Factors enhancing herbicide resistance include: the use of a single herbicide or herbicides of same mechanism of action, same formulation, same method of application, time of application, weather conditions during spraying, weed-density and application rate, surfactants, herbicide family and mechanism of action, crop rotation, and employed control methods.

Because weeds contain a tremendous amount of genetic variation that allows them to survive under a variety of environmental conditions, the development of a resistant species is brought about through selection pressure imposed by the continuous use of an herbicide or herbicides of similar mechanism of action. Long residual pre-emergence herbicides or repeated application of post-emergence herbicides will further increase selection pressure.

Factors in general that can lead to or accelerate the development of herbicide resistance include weed characteristics, chemical properties and cultural practices.

Weed characteristics conducive to rapid development of resistance to a particular herbicide include:

- Weeds having short life cycles (annuals).
- High seed production.
- Level of selection pressure imposed by the herbicide
- Relatively rapid turnover of the seed bank due to high percentage of seed germination each year (i.e., little seed dormancy).
- Several reproductive generations per growing season.
- Extreme susceptibility to a particular herbicide.
- One weed which would normally be controlled but not controlled while others were removed.
- High frequency of resistant gene (s).

Herbicide characteristics which lead to rapid development of herbicide resistance in weed biotypes include:

- A single site of action of the same herbicide continuously is used.
- Broad spectrum of weed control.
- Long residual activity in the soil.

Cultural practices can also increase the selection pressure for the development of herbicide-resistant biotypes. In general, complete reliance on herbicides for weed control can greatly enhance the occurrence of herbicide-resistant weeds. Other factors include:

- Shift from crop rotations towards mono cropping.
- Little cultivation or zero tillage for weed control or no elimination of weeds that escape herbicide control.
- Continuous or repeated use of a single herbicide or several herbicides that have the same mechanism of action.
- High herbicide use rate relative to the amount needed for weed control.
- Complete weed control
- Orchard and vineyard weeds.
- Roadside weeds.

MANAGEMENT OF HERBICIDE RESISTANCE

Herbicide-resistant weed populations can be managed following an integrated weed control program. The following practices are important for an effective management strategy:

- Herbicide rotation. Adopting this method, it should be known that herbicides of different chemical families may have the same site of action.
- Using mixtures of herbicides with different modes of action and overlapping weed spectrums. This would help in managing evolution of weed resistance.
- Crop rotation. Crops differ in their competitiveness against weeds. Plant crops having a different season of growth, different registered herbicides and crops for which there are alternate methods of weed control. Rotation breaks down weed population and prevents the build up of resistance to herbicides. In addition, different crops may require different types of herbicides and thus herbicides may be rotated as well. However, some herbicide groups include different chemicals that can be used in different crops; therefore crop rotation alone may not be enough to avoid resistance development in this case.
- Herbicides with the same site of action should not be applied or used in both fallow years and in the crop(s) planted within 3 years.
- Growers should keep rotating methods of weed control. Non-chemical control techniques including tillage, hand-weeding before flowering, mulching, soil solarization, prevention methods of weed dispersal (certified seed, clean equipments, use a power washer or compressed air to remove seeds).
- Herbicide-resistant weeds should be controlled before flowering and seed setting.
- Farmers should only use non- or short-residual herbicides and avoid using persistent chemicals and not applying them repeatedly within a growing season. This method would reduce the selection of herbicide-resistant weed biotypes. However, repeated applications within a single growing season of certain

herbicides (paraquat, glyphosate) also lead to development of resistant weed populations.

- Where possible mechanical weed control such as rotary hoeing and cultivation is recommended to be combined with herbicide treatments.
- Weed escapes of resistant biotypes may be eliminated by cultivation in row crops. Fallow tillage can control herbicide-resistant and susceptible weed populations when they emerge at about the same time.
- Accurate record keeping. Farmers should be familiar with the history of herbicides use in their fields. Also keep tracking the weed species that have been present in a given field and of how well particular herbicides have controlled them. Farmers should check for weedy patches in patterns consistent with application problems and hand-weeding these patches.
- Always weed free crop seeds should be used that greatly minimize introduction seeds of herbicide-resistant biotypes.
- Implementation of integrated weed management. This is important for effective control of all weeds including herbicide-resistance.
- Monitoring fields for weed escapes for resistant and susceptible biotypes. A resistance problem may not become visible until 30 percent or more of the weed population is no longer controlled. Check to see if the escapes are of one species or a mixture of species. If a mixture, the problem is more likely related to the environment or the herbicide application. If only one species was not controlled, the problem is likely to be resistance, especially if the species was controlled by the herbicide in the past and if the same herbicide has been used repeatedly in the field.
- Implementation of prevention methods of weed control. All measures aimed at prevention of weed introduction to fields and their dispersal should be strictly followed including governmental quarantine regulations.
- Alternating spring and winter crops, thus tillage and herbicides are used at different times in the different crops. Weed biotypes that survive in one crop could be killed in the other.
- Changing herbicide program, if weed resistance occurs, herbicides with other sites of action and other weed management practices

must be used in an integrated management strategy. However, weed management strategies that discourage the evolution of herbicide resistance should include the following:

Use herbicide only when necessary and where possible herbicide application should be based on economic threshold.

Apply herbicides in tank mixed, pre-packed, or sequential mixtures of multiple site of action.

Never use unregistered mixtures, follow label recommendation at all times

Regularly monitor your crops so that resistant patches can be observed in time to be controlled with, for instance, spot spraying.

Apply the herbicide at the correct leaf stage of the weed and the crop.

Calibrate sprayer correctly before using herbicides

Planting new herbicide-resistant crop varieties should not result in more than two consecutive applications of herbicides with the same site of action against the same weed unless other effective control practices are also included in the management system.

Respond quickly to changes in weed populations to restrict spread of weeds that may have been selected for resistance.

Encourage railroads, public utilities, highway departments and similar organizations that use total vegetation control programs and vegetation management systems that do not lead to selection of herbicide resistant weeds. Resistant weeds from total vegetation control areas frequently spread to cropland. Chemical companies, governmental agencies, and farm organizations can all help in this effort.

- To keep herbicide-resistant weeds under control, the following strategies should be also incorporated into a weed management plan:

Clean tillage and harvest equipments before moved from infested to clean fields from weed resistant species.

Total weed control in uncultivated places or sites

Close cultivation

Monitor hand weeding to insure more than 90% removal of weeds in the crop row.

Prevention of weed seed spread through:

- – Use of clean equipment.
- – Enter the field with resistant plants last.
- – Use a power washer or compressed air to remove seeds.
- – Monitoring the initial evolution of resistance by:
- – Recognizing patterns of weed escapes typical of resistant plants
- – Watch for small weed patches that appear in the same place in the next crop.
- – Watch for weed patches that do not have a regular shape that would indicate an herbicide application problem.
- Control suspected herbicide-resistant weeds before start seeding

Herbicide resistance however, provides a basic understanding of the genetic basis of weediness, while the development of weed genomics would provide three predictable and useful outcomes. The first is the identification of genes that could improve crop yields. The second is to improve our understanding of the evolution of herbicide resistance and the to aid in the identification of novel herbicide targets. Currently, there is little (if any) solid predictive capability of why some weeds develop resistance and others do not. Third, our understanding of weed biology would be exponentially expanded [6].

Research has recently been performed to assess the ability to cripple the effect of transgenes. The goal here is for the transgenic effect to not be as strong if it went to a wild relative. In one case, the genetic background of the crop weakened the weedy relative. In another case, the weakness was built into the genetic construct, called *transgenic mitigation*, in which an herbicide-resistant gene was paired with a dwarfing gene. In either case, transgenic weeds were less competitive than their non-transgenic parent weeds [39].

CONCLUSIONS

Weeds either leave (disappear), adapt, tolerate or resist any unfavorable environmental conditions that influence their normal growth and life strategies. Herbicide resistance is a complex phenomenon resulting from altered herbicide target enzyme, enhanced herbicide metabolism or reduced herbicide absorption and/or translocation. It is a survival strategy through which many successful weed species and/or biotypes counteract or escape chemical hazards. Weeds expressing this phenomenon have developed some morpho- (behaviorist), physio-, and/ or biochemical mechanism/s allowing existence. However, two theories are mainly considered: the mutation and the natural selection [17]. Colonizers, as well as some specialist weeds of high seed production and polymorphic characteristics, have rapid responses to prevailing environmental conditions and high ability to express herbicide-resistant genes and exhibit wide ecological variations [28]. This phenomenon is well documented in agricultural as well as other disturbed habitats while the list of weed resistant species gets longer with continued dependence on herbicides for weed control. From the information presented in this chapter, it is clearly demonstrated that herbicide resistance in weeds is far exceeding herbicide technology and industry. Most problematic weed species are genetically related to major food crops including wheat, rice and maize. This may pose another danger for the genetic industry and genetically engineered crops of wild relatives. Away from weed biology and resistance control, methods of weed control must be integrated and continuously rotated for effective weed control and prevention of weed resistance. This however, may not be achieved in absence of information and field data and well managed weed control strategies, considering all the factors that influence weed life and development.

REFRENCES

1. Anonymous-I. (2010) Evolution. Retrieved December, 2010. Available at: *http://www.biologydaily.com/biology/Evolution*.
2. Anonymous-II. (2010). Gene flow. Retrieved December, 2010. Available at: *http://en.wikipedia.org/wiki/Gene_flow*.
3. Anonymous-III. (2010) Gene flow mitigation. Retrieved December, 2010. Available at: *http://en.wikipedia.org/wiki/Gene_flow*.
4. Arnaud, J.F.; Viard, F.; Delescluse, M. and Cuguen, J. (2003) Evidence for gene flow via seed dispersal from crop to wild relatives in *Beta vulgaris*(Chenopodiaceae): consequences for the

release of genetically modified crop species with weedy lineages. *The Royal Society, Published online, 19 March, http://www.ncbi.nlm.nih.gov/pmc/articles.*

5. Barrett, S.C. (1988). Genetics and evolution of agricultural weeds. In: *Weed management in Agroecosystems: Ecological Approaches*. Altieri, M.A. and Liebman, M. (eds.). CRC Press, Inc, Baco Raton, Florida.
6. Basu, C.; Halfhill, M.D.; Mueller, T.C and Stewart Jr, C. N. (2004). Weed genomics: new tools to understand weed biology. *Tends in Plant Science*9(8), 391-398.
7. Campbell, L.G.; Snow, A.A and Ridley, C.E. (2006). Weed evolution after crop gene introgression: greater survival and fecundity of hybrids in a new environment. *Ecology Letters* 9, 1198–1209.
8. Chen, L.I.J., Lee, D.S., Song, Z.P., Shu, H.S. and Lu, B.R. (2004). Gene flow from cultivated rice (*Oryza sativa*) to its weedy and wild relatives.*Annals of Botany* 93, 67-73.
9. Conard, S.G. and Radosevich, S.R. (1979). Ecological fitness of *Senecio vulgaris* and *Amaranthus retroflexus* biotypes susceptible or resistant to atrazine. *Journal of Applied Ecology* 16, 171-177.
10. Duke, S.O.and Cerdeira, A.L. (2005). Transgenic herbicide-resistant crops. *Outlooks of Pest management.* DOI10.1564/16 oct06, pages 2008-211. Research Information Ltd.
11. DuPont (2008). Herbicide resistant crops and weed management: scientific summary and the DuPont perspective. Available at: http://www2.dupont.com/Biotechnology/en_US/science_knowledge/herbicide_resstance.
12. Ellstrand, N.C. (1992). Gene flow among seed plant populations. *New Forests Journal* 6(1-4), 241-256.
13. Ellstrand, N.C. (2003).Current knowledge of gene flow in plants: implications for transgene flow. *The Royal Society, published online 12 May at: http://www.ncbi.nlm.nih.gov/pmc/articles.*
14. Fuentes, R.G., Baltazar, A.M., Merca, F.E., Ismail, A.M. and Johnson, D.E. (2010). Morphological and physiological responses of lowland purple nutsedge (*Cyperus rotundus* L.) to flooding. *AoB Plants* Vol. 2010.plq010, doi: 10.1093/oobpla/plq010.

15. Gealy D.R.; Bradford K.J.; Hall L.; Hellmich R.; Raybould A. and Wolt J. (2007). Implications of gene flow in the scale-up and commercial use of biotechnology-derived crops: Economic and policy considerations. *Issue Paper 37. Council for Agricultural Science and Technology (CAST), Ames, Iowa.*
16. Gunsolus, J.L.(2008). Herbicide resistant weeds. 2002 Regents of the University of Minnesota.. University of Minnesota Extension. Available at:http://www.extension.umn.edu/distribution/cropsystems/DC6077.html.
17. Hager, A.G. and Refsell, D. (2008). Weed resistance to herbicides. In: *2008 Illinois Agricultural Pest Management Handbook*. Department of Crop Science, University of Illinois.
18. Herbicide Resistance Action Committee (HRAC). (2009). Guidelines to the management of herbicide resistance. Available at:http://www.hracglobal.com/Publications/ManagementofHerbicideResistance/tabid/225/Default.asps(1-16)1/18/2009.
19. Holm, L.G., Plucknett, D.L., Pancho, J.V. and Herberger, J.P. (1977). *The World's Worst Weeds: Distribution and Biology*. University of Hawaii, Honolulu. USA.
20. Holzner, W. and Numata, M. (eds.). (1982). *Biology and Ecology of Weeds*. Dr. W. Junk Publishers. The Hague-Boston-London.
21. http://www.weedscience.org/summary/UspeciesMOA.asp?1stMOAID=19.
22. Kending, A. (2009). Herbicide resistance in weeds. University of Missouri Extension. Available at: http://extension.missouri.edu/publications/DisplayPub.aspx?P=G4907.
23. LeBaron, H.M. and Gressel, J.(eds.). (1982). *Herbicide Resistance in Plants*. John Wiley & Sons, New York.
24. Lu, B-Rong and Yang, C. (2009). Gene flow from genetically modified rice to its wild relatives: assessing potential ecological consequences.*Biotechnology Advances* 27, 1083-1091.
25. Lu, B-Rong. (2008). Transgene escape from GM crops and potential biosafety consequences: an environmental perspective. International Center for Genetic Engineering and Biotechnology (ICGEB), vol. 4. Collection of Biosafety Reviews 66-141.

26. Lu, B-Rong. and Snow, A. (2005). Gene flow from genetically modified rice and its environmental consequences. *BioScience,* 55, 669–678.

27. Mashburn S. J. Sharitz, R. R. and Smith, M. H. (1978). Genetic variation among *Typha* populations of the southeastern United States. *Evolution* 32: 681-685.

28. Neve, P. (2007). Challenges for herbicide resistance evolution management: 50 years after Harper. *Weed Research* 47, 365-369.

29. Putwain, P.D., Scott, K.R. and Holliday, R.J. (1982). The nature of resistance to traizine herbicides: case histories of phenology and population studies. In: *Herbicide Resistance in Plants;* LeBaron, H.M. and Gressel, J. (eds.), John Wiley & Sons, New York pp. 99-115.

30. Qasem, J. R. (2003). *Weeds and their Control.* University of Jordan Publications. Amman, Jordan. 628 pp.

31. Qasem, J. R. (2011). Herbicides applications: Problems and considerations, In, *Herbicides and Environment,* Andreas Kortekamp (Ed.), *ISBN*: 978-953-307-476-4, InTech, Available from: http://www.intechopen.com/articles/show/title/herbicides-applications-problems-and-considerations.

32. Qasem, J. R. and Al-Jebury, I. S. (2001). Weed control in marjoram (*Origanum syriacum* L.) under field conditions. Dirasat 28, 194-207.

33. Qasem, J.R. and Foy, C.L. (2006). Selective weed control in Syrian marjoram (*Origanum syriacum*L.) with oxadiazon and oxyfluorfen herbicides.*Weed Technology,* 20 (3), 670-676.

34. Quist D. (2010). Vertical (Trans) gene Flow: Implications for crop diversity and wild relatives published. *Third World Network, Penang, Malaysia Jutaprint, ISBN:* 978-967-5412-26-4.

35. Radosevich, S. R., Holt, J.S. and Ghersa, C.M. (2007). *Ecology of Weeds and Invasive Plants. Relationship to Agriculture and Natural Resource Management.* Hobokin, N.J.: Willey Interscience.

36. Radosevich, S., Holt, J. and Ghersa, C. (1997). *Weed Ecology: Implication for Management* . 2nd Edition. John Wiley & Sons Inc. New York.

37. Smith, C.M. and Namuth, D. (2005). Herbicide resistance: mechanisms, inheritance, and molecular genetics. eLearn &

Grow Library, 7pp. available at: http://plantandsoil.unl.edu/croptechnology2005/pageincludes/printModule.jsp?inform.

38. Smith, C.M.; Hulting, A.; Thill, D.; Morishita, D. and Krenz, J. (2007) Herbicide-resistant weeds and their management, weed control strategies. In:*Pacific Northwest Conservation Tillage Handbook, University of Idaho.*

39. Stewart N.C. (2008). Gene flow and the risk of transgene spread. University of Tennessee. Retrieved December 2010 Available at: http://agribiotech.info/details/Stewart-GeneFlow%20Mar%208%20-%2003.pdf.

40. Tsyusko, O. V., Smith, M. H., Sharitz, R. R. and Glenn, T. C. (2010). Genetic and clonal diversity of two cattail species, *Typha latifolia* and *T. angustifolia* (Typhaceae), from Ukraine. *American Journal of Botany,* 97 (12), 2061-2067.

41. van de Wiel, C.; Groot, M.; and Den Nijs, H. (2005). Gene flow from crops to wild plants and its population ecological consequences in the context of GM-crop biosafety including some recent experiences from lettuce. *Wageningen UR Frontis Series, Volume 7 Environmental Costs and Benefits of Transgenic Crops.* Retrieved, December 2010 Available at: http://library.wur.nl/ojs/index.php/frontis/article/viewArticle/914.

42. Vencill, W., Grey, T. and Culpepper, S. (2011). Resistance of weed to herbicides. In: *Herbicides and Environment* (Kortekamp, A. ed.). pp. 585-594. InTech, Available at: http://www.intechopen.com/articles/show/title/herbicides-applications-problems-and-considerations.

43. White, D.E. (1979). *Physiological Adaptations in Two Ecotypes of Canada Thistle (Cirsium arvense* (L.) Scope. MSc. Thesis, University of California, Davis.

44. WSSA, (2012). Data from the international survey of herbicide resistant weeds obtainable at WeedScience.com.website.

Citations

CHAPTER 1

B. Samedani, A. S. Juraimi, M. P. Anwar, M. Y. Rafii, S. H. Sheikh Awadz, and A. R. Anuar, "Competitive Interaction of Axonopus compressus and Asystasia gangetica under Contrasting Sunlight Intensity," doi:10.1155/2013/308646.

CHAPTER 2

John Carter, Allison Jones, Mary O'Brien, Jonathan Ratner, and George Wuerthner, "Holistic Management: Misinformation on the Science of Grazed Ecosystems," International Journal of Biodiversity, vol. 2014, Article ID 163431, 10 pages, 2014. doi:10.1155/2014/163431.

CHAPTER 3

F. Stuart Chapin III, Erika S. Zavaleta, Valerie T. Eviner, Rosamond L. Naylor, Peter M. Vitousek, Heather L. Reynolds, David U. Hooper, Sandra Lavorel, Osvaldo E. Sala, Sarah E. Hobbie, Michelle C. Mack, and Sandra Díaz, Consequences of changing biodiversity, doi:10.1038/35012241.

CHAPTER 4

King N, Biggs H, Loon R. Seeking Common Ground: How Natural and Social Scientists Might Jointly Create an Overlapping Worldview for Sustainable Livelihoods: A South African Perspective. Conservat Soc 2007;5:88-114.

CHAPTER 5

Amelia T. Elgar, Kylie Freebody, Catherine, L. Pohlman, Luke P. Shoo, and Carla, P. Catterall, Overcoming Barriers To Seedling Regeneration During Forest Restoration On Tropical Pasture Land And The Potential Value Of Woody Weeds , doi: 10.3389/fpls.2014.00200.

CHAPTER 6

Patrik Kehrli, and Steve D. Wratten, "A Perspective on the Consequences for Insect Herbivores and Their Natural Enemies When They Share Plant Resources," ISRN Ecology, vol. 2011, Article ID 480195, 6 pages, 2011. doi:10.5402/2011/480195.

CHAPTER 7

Lahoz-Monfort JJ, Guillera-Arroita G, and Hauser CE (2014) From planning to implementation: explaining connections between adaptive management and population models. Front. Ecol. Evol. 2:60. doi: 10.3389/fevo.2014.00060.

CHAPTER 8

John Derek Scasta, David M Engle, Ryan N Harr, and Diane M Debinski, Fire Induced Reproductive Mechanisms of a Symphoricarpos(Caprifoliaceae) Shrub After Dormant Season Burning, doi:10.1186/s40529-014-0080-4.

CHAPTER 9

Zina Skandrani, Sébastien Lepetz, and Anne-Caroline Prévot-Julliard, Nuisance species: beyond the ecological perspective, doi: 10.1186/2192-1709-3-3.

CHAPTER 10

Corinna Riginos, Lauren M Porensky, Kari E Veblen, Wilfred O Odadi, Ryan L Sensenig, Duncan Kimuyu, Felicia Keesing, Marit L Wilkerson, and Truman P Young, Lessons on the Relationship between Livestock Husbandry And Biodiversity From The Kenya Long-Term Exclosure Experiment (KLEE), doi:10.1186/2041-7136-2-10.

CHAPTER 11

Jamal R. Qasem, Herbicide Resistant Weeds: The Technology and Weed Management, DOI: 10.5772/56036.

Index

D

E

F

G

H

I

K

L

M

N

O

P

R

S

T

U

W